Alternative Pathways in Science and Industry

Urban and Industrial Environments
Series editor: Robert Gottlieb, Henry R. Luce Professor of Urban and Environmental Policy, Occidental College

Alternative Pathways in Science and Industry

Activism, Innovation, and the Environment in an
Era of Globalization

David J. Hess

The MIT Press
OCM 71507283
Cambridge, Massachusetts
London, England

For information on quantity discounts, email special_sales@mitpress.mit.edu.

Set in Sabon by The MIT Press. Printed and bound in the United States of America.

Library of Congress Cataloging-in-Publication Data

Hess, David J.
Pathways in science and industry : activism, innovation, and the environment in an era of globalization / David J. Hess.
   p.   cm. — (Urban and industrial environments)
Includes bibliographical references and index.
ISBN-13: 978-0-262-08359-1 (hardcover : alk. paper)
ISBN-13: 978-0-262-58272-8 (pbk. : alk. paper)
1. Science—Social aspects.   2. Technology—Social aspects.   3. Science and industry. I. Title.
Q175.5.H469 2007
303.48'3—dc22

2006030374

10   9   8   7   6   5   4   3   2   1

Contents

Acknowledgments

My long-term affiliation with the Department of Science and Technology Studies at Rensselaer Polytechnic Institute and the remarkable interdisciplinary conversations that have occurred there over the years are, without a doubt, my first debt. I mention in particular the many conversations over the years with my colleague Langdon Winner and our collaboration on the National Science Foundation grant "Sustainable Technology, the Politics of Design, and Localism" (SES-00425039). The grant gave me extra time that was helpful for finishing this book, and it also allowed me to work on a sibling project with Langdon on the intersections of sustainability projects with localism, democracy, and justice. Four graduate students (Richard Arias-Hernández, Colin Beech, Rachel Dowty, and Govind Gopakumar) also participated in the grant, and their specific contributions are noted in the by-lines of the case studies available on the project's website (http://www.davidjhess.org). I have also benefited from conversations over the years with Sal Restivo about science studies; Kate Boyer, Frances Bronet, Linnda Caporael, Ron Eglash, Ray Fouche, Gary Gabriele, Ted Krueger, Mark Mistur, Michael Oatman, Mark Steiner, Dan Walczyk, and Ken Warriner about design; Steve Breyman, Carl McDaniel, and Ned Woodhouse about technology and the environment; and Nancy Campbell about the welfare state. I also appreciate the opportunity to have worked with Steve Breyman, Nancy Campbell, and Brian Martin on a related review essay titled "Science, Technology, and Social Movements," which is forthcoming in the *Handbook of Science and Technology Studies* (MIT Press). Additional support to the department and the university comes from two National Science Foundation grants: "An STS Focus on Design" (SES 9818207) and "Product Design and Innovation" (CCLI 9950931).

Of the many colleagues in science and technology studies and related fields, I mention in particular recent conversations with and work by Julian Agyeman, Phil Brown, Maurie Cohen, Arturo Escobar, Frank Fischer, Scott Frickel, Ken Gould, Sandra Harding, Krista Harper, Andy Jamison, Ulrik Jørgensen, Daniel Kleinman, Brian Martin, Arthur Mol, Kelly Moore, Steven Moore, and Roddey Reid, who share similar interests in social movements, the environment, and/or science and technology. Much earlier in my career, Davydd Greenwood introduced me to Weber's *The City*, Jim Boon and David Holmberg opened the door to Weber's comparative religious studies, and Roberto DaMatta taught me a Brazilian structuralist reading of Weber. Tom Holloway introduced me to the comparative literature on Latin America. They have left an imprint that will be evident in the comparativist methodology that I continue to use.

Numerous conversations with Rensselaer's former and current doctoral students and visiting students—and the privilege of accompanying and sharing in their dissertations, graduate study, and careers—influenced this volume. They include, in addition to the students mentioned above, Barbara Allen, Sulfikar Amir, Dikoh Chen, Ayala Cnaan, Seval Dulgeroglu, Maral Erol, Virginia Eubanks, Nicole Farkus, Patrick Feng, James Fenimore, Jenrose Fitzgerald, Jill Fisher, Ken Fleischmann, Noelle Foster, Jaime Radesi Galayda, Andy Karvonen, Natasha Lettis, David Levinger, Lisa McLoughlin, Torin Monahan, Rosana Monteiro, Casey O'Donnell, Tolu Odumosu, Steve Pierce, Hector Postigo, Lorna Ronald, Emily Valerio, Roli Varma, Electra Weeks, and Margaret Wooddell. I have pointed to their work in the notes. Various sections benefited from suggestions and conversations with people with expertise in the area, including Jason Patton, Kate Boyer, Michael Buser, and Jeff Hannigan on New Urbanism and Melanie Dupuis and Neiva Hassanein on agriculture. I also appreciate the comments from faculty and students in the seminars given at the Center for Agroecology and Sustainable Food Systems at the University of California at Santa Cruz and the Center for Sustainable Development at the University of Texas at Austin. I am also grateful for the help and guidance of my editor at The MIT Press, Clay Morgan.

An earlier work containing material that now appears in chapters 1–3 was published electronically on my website, with a few print copies circulated to colleagues and libraries (Hess 2001c). The experiment in electronic publication brought the luxury and curse of allowing me to revise, and the end result is the present book. There may be also some limited overlap in content with the following publications:

"Technology- and Product-Oriented Movements: Approximating Social Movement Studies and STS," *Science, Technology, and Human Values* 30 (2005), no. 4: 515–535

"Medical Modernization, Scientific Research Fields, and the Movement for Complementary and Alternative Cancer Therapies," *Sociology of Health and Illness* 26 (2004), no. 6: 695–709

"Object Conflicts in a Health-Environmental Social Movement: The Movement for Organic Food and Agriculture in the U.S.," *Science as Culture* 13 (2004), no. 4: 493–514.

In all cases contractual agreements provide permission for republication of portions of the previously published work in this book.

Finally, I wish to thank Gretchen for helping me to juggle the time that has made writing this book possible.

Introduction

The Honest Weight Food Coop in Albany, New York, serves both middle-class shoppers who drive in from miles around and the ethnically and economically mixed residents of the urban neighborhood in which the Coop is located. In addition to providing the region and the neighborhood with high-quality, reasonably priced organic food from local farms and distant cooperatives, the Coop is a democratically run organization that serves as a node in many of the region's alternative networks. However, it is also a store that must compete in a retail marketplace. Although it has a loyal following and is a treasured local institution, like other retail food cooperatives across the United States it is potentially threatened by changes in the supermarket retail industry. At the time of this writing, the Coop does not have direct competition from the growing chains of natural foods stores; however, regional supermarkets have developed store-within-a-store departments that provide many of the same products that the Coop offers. The supermarket has literally incorporated the free-standing natural foods store into its retail space.

Although the supermarket has an increasing overlap of products, it has also made some significant changes. For example, the cooperatives' emphasis on produce grown by local farmers has given way to an emphasis on organic foods trucked in from distant farms, and there is relatively little emphasis on fresh organic foods relative to processed offerings. Although my family prefers the Coop, we also spend a portion of our monthly food budget at the nearby supermarket, which has a natural foods section. Not only is the supermarket much closer; it also has many items that are not available at the Coop. Life is full of compromises, and this is a book about compromises.

The dynamic of the cooperative and the supermarket brings to life many of the themes and problems that this book analyzes from a theoretical and synthetic perspective for fields such as renewable energy, recycling and remanufacturing, and "green" infrastructure and building design. First, organic food is an example of the many technological innovations that have emerged largely from the grassroots of environmentally oriented social movements and associated entrepreneurs. Behind the alternative product is an alternative knowledge and technology that had to be developed in order to bring it into existence. Two questions that grassroots innovation opens up are "What roles have social movements and activists played in scientific, technological, and industrial innovation?" and "How can scientists, designers, and entrepreneurs be viewed as simultaneously epistemic, economic, and political actors?" Too often, we think of environmental activism and social movements as merely opposing scientific research and industrial innovation (such as nuclear energy and genetically modified foods), but there is another, generative side of social movements that occurs along with industrial opposition.[1]

A second major issue that the case of the cooperative and the supermarket crystallizes is the way that dominant economic institutions, in this case supermarkets and the food-processing industry, have both absorbed the alternative products and transformed them. The example suggests that although mainstream industries may at first ignore or even discourage the grassroots innovations, at some point they tend to take up the challenges and rework them. However, the incorporation of the alternatives into their own product lines, supply chains, and retail spaces is not a straightforward process; instead, incorporation also tends to coincide with the transformation of both alternative ownership structures and the material qualities and design of the alternative products and technologies. The process of incorporation and transformation is another example of how the issue of environmentally oriented social movements and their relationship to science and industry should be reexamined.[2]

To explore the complex dance of incorporation and transformation, of compromises on both sides, and in the process to understand better the complex politics and histories of environmentally oriented industrial innovation, this book draws on and modifies work developed in social studies of science and technology, social movements, and globalization.

The objective of this book is largely theoretical: to suggest ways in which the three related research fields can be brought together and moved forward.

I argue that researchers, activists, and innovators should pay more attention to three kinds of historical change. First, the research agendas of science and technology have become increasingly open to the scrutiny and influence of industrial funders as well as grassroots consumer and non-profit groups. A theory of scientific and technological change should accommodate the roles of both industry and civil society, and consequently the theory should build on but also move beyond the constructivist models of science and technology that have dominated studies of science and technology. Second, social movements have increasingly broadened the target of mobilization from governments to industrial corporations, and they have also broadened their work from opposition to creative reconstruction. A theory of social movements should accommodate the ways in which the values of social movements have reached into entrepreneurship, consumption, design, industrial innovation, and even hobbies and charitable activities. Finally, the institutional changes associated with globalization have generated a countervailing localization of economics and politicization of consumption, where social movements find new challenges and opportunities. A theory of globalization should recognize the growth of counter-globalization economics that are carried out through the politics of design, innovation, consumption, and localism.

All three of the changes have been noted in the respective literatures; the goal of this project is to explore their intersections and reveal something new in each. In the process, the book will develop the concept of "alternative pathways" and explore both their limitations and their potential for bringing about social change.

Social Movements and Alternative Pathways

One of the changes associated with the era of globalization is the shift of social movements away from repertoires based on strikes and street protest, on direct political action by political groups against governments and unions against large corporations, and on a goal of attaining greater economic and political rights for the poor, the working class, and the oppressed in general. Instead, the literature has recognized that social movements can also include repertoires of lifestyle change and economic activity, action in favor of building alternative institutions, and a goal of change in cultural practices. Much of the discussion of a broader view of social movements has taken place under the rubric of "new social movements." Rather than debate the value of the concept or linger on contrasts with "old" social movements (such as the labor movement), I focus on the generative influence of social-movement action and related action on scientific, technological, and industrial innovation.[3] To examine the question, I have found it necessary to avoid prematurely restricting the scope of analysis to a narrow definition of social movements. The perspective is consistent with the social-science literature that has submerged the concept of a social movement in the broader category of contentious politics. However, because the comparative field of analysis that I want to examine includes the creation of alternative businesses, household activities, and non-profit organizations, the term "contentious politics" also may be overly restrictive in some cases. For this reason, I have used the concept of an "alternative pathway."[4]

The concept of an alternative pathway makes it possible to relax the analytical requirement that would exclude from consideration types of social action that do not meet a definition of social movements. It allows the inclusion of organizations that have social-change goals as well as organizations that do not have an explicit or self-conscious goal of fundamentally changing society. For example, some of the alternative pathways exhibit complex mixes of social-change goals with goals of profitability, faith-based charity, or even leisure-time hobbies such as gardening and home tinkering. Although some organizations may mix social-change goals with other goals, they may not view themselves as engaging in contentious politics, and they may not see themselves as belonging to a social movement. Furthermore, organizations that were originally dedicated to social change sometimes shift their goals. The concept of an "alternative pathway," at least as I will use it here, will make it possible to avoid drawing premature boundaries when confronted with the fluidity of goals and repertoires of action.[5]

However, because alternative pathways often take the form of social movements, theories of social movements provide a good starting point for the study of alternative pathways. There is no single, widely accepted definition of a social movement, and in any case definitions are conventions that delineate a field of comparative analysis and focus attention on some problem areas and research projects as more worthy of attention than others. Some would prefer to restrict the scope of the definition of a social movement to protest-based social-change action that attempts to defend the rights of ethnic minorities, colonized peoples, women, working people, and other historically oppressed groups. Such a perspective is neither right nor wrong in a scholarly sense; it merely directs attention toward some issues and away from others.

One aspect of social movements that will be important for the study of alternative pathways is that they are located in civil society, that is, a sector of society that can be distinguished from the governmental, private, and domestic sectors. I will understand the crucial distinctive feature of civil society to be the use of voluntary donations of labor and/or funds as the means for organizational support and reproduction. The definitional focus on voluntary action recognizes that civil society includes a wide variety of organizations other than social-movement organizations, including religious, charitable, sporting, leisure, professional, hobby, and ethnic organizations. Although I will understand voluntary action as the primary distinctive feature of civil-society organizations, such organizations can often be found doing work that is associated with governments, firms, and households. In other words, they sometimes monitor and contribute to the regulation of society, produce goods and services for sale on the market, and provide, for example, care to children and assistance to the elderly. The overlap of some civil-society organizations with the development and production of goods and services will be of particular interest in the study of the alternative pathways. Because private-sector firms, government agencies, and in some cases alternative living arrangements appear in some alternative pathways, the concept is not limited to civil-society organizations.

Social-movement organizations are a type of civil-society organization that is distinguished by a unique cluster of features. The following definitions of the features are arguably well within the range of current usages of the term in the broader literature on social movements:

1. broad scope in terms of organizational diversity and temporal duration
2. a goal of bringing about fundamental social change from groups that are disempowered or perceive themselves to be disempowered on at least some issues

3. repertoires of action that include the use of extra-institutional strategies such as protest.

The first criterion appears to be the simplest but may be the most difficult to evaluate. A social movement can be distinguished from a single organization or a small network that engages in activism, and it can be distinguished from a single campaign that is restricted in topical and temporal scope. Ultimately, distinctions between a movement and an activist network, or between a movement and a campaign, are not precise, but they can be helpful as guides for selecting points of comparison and using a terminology that has at least some boundaries.[6]

The second criterion has considerable variability. The often-used phrase "promote or oppose social change" tends to narrow the comparative field in comparison with alternatives such as "challenge to authority" and "articulates a social conflict." The latter phrases are useful for some purposes, but they open up the field of inquiry to include topics such as deviance and crime, and consequently they are unnecessarily broad for the present purposes. Likewise, the emphasis on a goal of social change also helps to distinguish social-movement organizations from interest groups and charitable groups, which seek to redistribute resources but do not seek fundamental social change. Although the distinction between a goal of fundamental social change and a shift in the allocation of resources is fuzzy, I have found it helpful especially for the discussion of the type of the alternative pathway that advocates increased access for the poor.[7]

The third component of the definition, extra-institutional strategy, usually coincides with boundary work between those who study protest movements and those who study other types of multi-organizational, multi-campaign social-change action. It is useful to distinguish organizations and networks that engage in extra-institutional repertoires of action from those that do not, but the scope of analysis for this project is broader than a repertoire of protest strategies. Because the argument here is that it is valuable to analyze together various types of social-change action, a variety of terms is necessary. For example, I have found it useful to make a distinction between social movements and reform movements. Like social movements, reform movements are long-term, multi-organizational efforts to change society, but they operate through existing institutional channels. They may include networks of reformers and reform organizations within a profession, a scientific field, or an industry. Just as activists can coalesce into a social movement, individual advocates within governmental, business, civil-society, or scientific organizations can also coalesce into a more substantial reform movement. For social-change-oriented activity on a smaller scale, the terms "activism" and "advocacy" are used, respectively, for action that either does or does not involve extra-institutional strategies, especially protest. Likewise, the conventional term "interest group" is used for action that is not oriented toward fundamental social change and occurs within existing institutions. (See table I.1.)

The relationship between reform movements and social movements is often characterized by ambiguity and ambivalence. The work of reform movements can be viewed alternately as part of the cooptation strategy of elites, which may opt to divide and conquer a social movement by supporting a reform movement, or as expressions of divisions in elites that afford positive political opportunities for the mobilizing efforts of social movements. In the cases that I am examining, organizations that approximate the ideal types of social movements and reform movements often function together, but tensions may also develop. For example, social-movement organizations often try to recruit reformist scientist experts to their side, but the relationships with scientists are often permeated with ambivalence on both sides.[8]

Finally, there is some value in maintaining a distinction among reform movements, new industries, and charitable groups. They all work within institutional channels to achieve change, but they achieve different types of change. Reform movements are concerned with broad social change, but they work by attempting to reform a profession, a policy field, or an

Table I.1
Social movements and related categories.

Type of action	Scope	Social-change goal	Extra-institutional repertoire
Social movement	+	+	+
Reform movement	+	+	−
Activist network	−	+	+
Advocacy network	−	+	−
Riot, panic	+	−	+
Crime, deviance	−	−	+
Interest group	+/−	−	−

industry. Reform work can involve creating new professional organizations, non-governmental organizations (NGOs), or firms. New industries achieve technological and economic change, but generally with a primary goal of profit making rather than social change. However, there can be contributions from social movements to reform movements within industries that eventually create new industries. Charitable organizations that do not have social-change goals are an interest group for the poor or the needy, but some charitable organizations also advocate for broader social change and may see themselves as belonging to a broader social or reform movement, such as the anti-hunger movement. Although sometimes excluded from the literature on social movements as outside its purview, in this book charitable organizations will play a more significant role, partly because they sometimes result from a mission shift of former movement organizations and partly because they have developed important relationships with the other types of alternative pathways.

Environmental Crises and Globalization

In the title of this book, I use the preposition "in" rather than "to." I view the pathways for societal change as elements of modernity rather than as anti-modern. From a long-term perspective, the study of globalization is a twenty-first-century version of the broader social-science problem of understanding modernity—that is, developing a historical theory of the fundamental processes of societal change during the last 500 years. To begin the discussion of globalization, it is necessary first to explain how I use the concept of modernity.

The founders of modern social theory each emphasized a different dimension of modern society that can provide a valuable, if limited, perspective on globalization today. I have occasionally drawn concepts from each of the three major traditions of social theory: from the Weberian and Parsonian tradition, the increasing universalism of standards and values (the increasing pervasiveness of normative systems that apply to all actors rather than to particular social groups); from the Durkheimian tradition, the increase in societal differentiation and the challenge of societal integration; and from the Marxist and world systems tradition, the expansion of scale of polities and economies based on accumulation processes and colonization of smaller-scale societies.[9] Although each of these three traditions offers a valuable approach to understanding globalization, each also tends to under-conceptualize the society-environment relationship, and it has become increasingly evident that there is a need to add a fourth major perspective on the study of modernity, which I term "the denaturalization of the world." The increasing mediation of nature-culture relations by science and technology has brought many benefits, but it has not been as uniformly beneficial as previous generations had anticipated. One of the defining characteristics of recent history is the collapse of the optimism associated with scientific progress and technological innovation. As Ulrich Beck has argued, during the course of the twentieth century the belief that advances in science and technology would lead inevitably to general societal progress broke down, and with the loss of innocence a modernist era of modern history drew to a close. Although concerns with science and technology were voiced in previous centuries, the various hazards and perceptions of the risks associated with modern technology (such as concerns with weapons of mass destruction, global warming, genetically modified food, computer-based surveillance, and genetic discrimination) created a crisis of confidence in the belief that scientific and technological advances were necessarily linked to progress.[10]

To the extent that there was a silver lining, the transformation in culture included recognition that society creates its environment as much as it adapts to it, and that technology has changed the environment into an increasingly synthetic product of human activity. Perhaps there is no better example of denaturalization than the looming uncertainty regarding the effects of human activity on the global climate. General understanding of the society-environment interaction, as well as concern with various environmental crises, has increased during the current phase of modernity, its "globalization" era. We appear to have created a globalized economic system that the global ecosystem, not to mention our own bodies, is incapable of supporting. Whether one focuses on climate change, on persistent chemical pollutants throughout the biosphere, on the impending scarcity of petroleum and natural gas, or on the relentless destruction of habitats and species, the environmental crisis is of global proportions. Global political and economic institutions have to date proved incapable of remediating the crisis, and in the wake of

their failure alternative pathways oriented toward environmental amelioration have proliferated. This is not to say that many other pressing
problems and social movements deserve attention, but only that the
focus in this book will be on the environmental crisis and the alternative
pathways in science and industry that have emerged to address its
various manifestations.[11]

My choice of the term "globalization" to describe the current historical period requires some explanation, because there are several
competing terms (e.g., "late modernity," "late capitalism," "postmodern
society"), and the choice of terms tends to indicate allegiances to specific
intellectual traditions. I have chosen "globalization" because it appears
to be relatively neutral among currently used terms, and it also draws
attention to the importance of ways in which societal problems, such as
the environmental crisis, are increasingly also global in scope. I avoid the
term "postmodern" because I assume that there has been no fundamental break with the dominance of the economic sector and because the
main institutions of modernity were present, at least in incipient forms,
as long ago as the sixteenth century, and in some cases earlier. We may
be in a postmodernist era in the sense of having left behind a naive faith
in scientific and technological progress that was characteristic of the late
nineteenth century and the early to middle years of the twentieth, but we
have not entered a postmodern era.

Because the concept of globalization will be a significant theoretical
point of reference in this book, it is necessary at the outset to define the
term clearly and to leave some excess baggage behind. For example, the
theses of the demise of the nation-state, the national economy, and the
welfare state should be inspected carefully and skeptically against economic data and historical case studies, both of which suggest
considerable comparative specificity. Furthermore, the belief that the
growth of international governmental organizations will coincide with
electoral democracy and accountability is questionable, in view of the
democracy deficit of many international institutions. Likewise, the belief
that the spread of multinational corporations and global media coverage
imply cultural homogenization has little basis in ethnographic evidence.[12]

One useful test for a list of distinctive features of globalization is to
compare the features posited for the late twentieth century and the early
years of the twenty-first with those posited for the modernist period (i.e.,

the years from the late nineteenth century to the middle of the twentieth), as Paul Hirst and Graham Thompson have done. Their approach has the advantage of leaving the concept of "globalization" open to empirical research that assesses the significant similarities and differences between present-day institutions and cultures in comparison with those of the earlier period. The cautious approach to the topic can help the researcher to avoid conflating a social scientific and historical analysis of globalization with ideological statements that draw overly sharp distinctions between a globalization era and some previous era (such as a Cold War or pre-World War II era), or that confuse the diffusion of new technologies and institutions with their democratization. The approach can also help clarify significant changes that have occurred.[13]

As a topic of empirical research, globalization can be divided into political, economic, and social dimensions. Regarding government and governance, rather than begin by accepting uncritically the thesis of the demise of the nation-state, it is better to frame the question as how, and under what conditions, the sovereignty of nation-states is and is not changing. Clearly, states share an increasingly crowded international stage with new and old types of NGOs, intergovernmental organizations, and transnational corporations. The changes do not mean that the era of domination of the international system by powerful states has ended. Although the Bretton Woods institutions have limited the sovereignty of poorer and indebted countries, the international financial institutions continue to be dominated by the large and wealthy states. Furthermore, control over military forces remains in the hands of the large and powerful states, which continue to dominate the international system through use (or threatened use) of military force.[14]

The shifts in sovereignty occur around the edges of the fundamental reality of state-based warfare—a reality that is continuous with earlier periods of history. The rise of non-state military actors such as terrorist networks may have eroded the confidence and power of states, but it has not ended their fundamental military dominance. Likewise, the growth of regional trading blocks and of the Organization of Petroleum Exporting Countries (OPEC) has entailed shifts of sovereignty for participating states, and at the international level the wealthy, industrialized nations contend with the shift in economic power—and, increasingly, political power—to India and China. There is also a shift downward toward the increased role of subnational regional economies and global cities and outward from states to NGOs and private firms. Together, those changes form the basis for an empirically grounded analysis of globalization as a political process.[15]

Of special interest for this book is another aspect of the changing role of the nation-state: the privatization of state functions under the ideological guidance of neoliberalism. In many scientific research fields the funding source has become more differentiated, with private foundations and private corporations increasingly prominent. In some fields research oriented toward technological innovation is now driven more by corporate research departments than by university laboratories. Even where university laboratories are a driving force, intellectual property agreements link the research to corporate ownership. A second aspect of privatization has been the shift of welfare state functions to the non-profit sector and to local governments. As I will explore in more detail in chapter 6, the privatization and devolution of welfare has had a tremendous influence on the historical trajectory of access-oriented alternative pathways.

In the economic sector, globalization is often characterized by the increased interconnectedness of the global economy since World War II. However, the trend follows a period of decentralization that can be traced back to World War I, and, as Hirst and Thompson have argued, by some measures the global economy was more integrated at the beginning of the twentieth century. Nevertheless, there are various ways to characterize the current levels of global economic integration that emphasize crucial differences from the modernist period. For example, although trans-oceanic cables connected financial markets as early as the 1860s, the growth of financial markets relative to gross domestic product, the digitalization of financial information, and the rise of Asian and Southern financial markets represent qualitative changes. Likewise, although large corporations have played a significant role in the international economy for centuries, in many industries across the world there has been a dramatic consolidation of ownership in the hands of a small number of large corporations. Companies that once purchased materials from local and regional sources and sold them in local or regional markets have shifted their strategy toward global sources and markets, or they have sold out to companies that have such a strategy. With the increasing prominence of global corporations and governance institutions, social movements have also become more transnational.[16]

In addition to the growth in financial markets and corporate consolidation, a third area of significant economic change involves income inequality. To the delight of neoliberal globalists, overall between-nation income inequality has declined since 1960. However, the effect is due largely to industrialization in Asia and some other regions, whereas sub-Sahara Africa has not kept pace with growth elsewhere. Another measure of inequality, the gap in per capita income between the richest one-fifth and the poorest one-fifth of the world's population, nearly doubled in the period 1970–1990. Within-nation inequality has generally increased, and the United States is no exception.[17]

Economists have not developed a consensus explanation for increased inequality in the United States, but one likely mechanism is the rise of corporate power as a result of globalization. During the 1980s union membership in the private sector declined by about ten points to 12 percent, and in manufacturing industries about 70 percent of firms undergoing unionization elections threatened to close their factories. Unionization rates have been double in industries such as health care, where the threat of capital flight is less powerful than in manufacturing. As of 2004, about 400,000 jobs were being shifted overseas per year, double the rate of just three years earlier. Outsourcing included an increasingly diverse group of industries, proportionately higher levels of unionized workers, and increasing numbers of the highly skilled occupations such as engineering and computer programming. The loss of jobs to low-wage overseas markets, the cutbacks in government aid programs, and the displacement of locally owned manufacturing and retail businesses by large corporations with distant headquarters are all aspects of a pattern of industrial concentration and globalization that has led to both unemployment and underemployment in many American cities. Those economic processes provide an important background for understanding the development of alternative pathways, especially the counter-globalization strategies of localization and the growth of access-oriented advocacy.[18]

Global competition has increased pressure on city, state, and national governments to encourage firms to invest in research and development for industries in which well-paid, highly skilled workers can be an asset.

The processes of adaptation include not only shifts toward post-Fordist production and ongoing product innovation but also increased reliance on technology transfer from university, government, and corporate research laboratories. Investments in science, technology, and innovation have become increasingly important survival strategies for the wealthy countries. National and local governments alike have backed the development of regional industry clusters that are interwoven with government and university partnerships. The advantages of regional clusters, in turn, strengthen a new form of local power that is oriented toward global markets. However, there is also a reactive process of import substitution, of attempts by cities destroyed by runaway shops to regain economic control at a local level by shifting consumer spending toward local products, especially in the agricultural, energy, and retail fields.[19]

In the case of the third dimension of globalization, increasing immigration and the question of cultural homogenization, it is also necessary to leave some common assumptions at the door. For example, in the United States there have long been significant flows of immigrants, including high levels during the last years of the nineteenth century and the early years of the twentieth. To understand immigration in the era of globalization, we should focus less on absolute numbers than on the forms of immigration. For example, since the early twentieth century there has been a shift from the emigration of Europeans to other world regions to the immigration of people from poorer, non-European countries to Europe and North America. In many cases, ethnic tensions that have always been associated with new immigrants have been exacerbated by religious differences and by the precarious citizenship status that some of the new immigrant groups have. Furthermore, transportation and communication technologies that were unavailable or more costly in the nineteenth and early twentieth centuries have made it easier for new immigrant groups to maintain diasporic links and to resist assimilation policies aimed at producing national identity and linguistic uniformity. The ethnographies on diasporic and ethnic minority communities provide sufficient data to counter facile claims that cultural homogenization is an inevitable result of globalization; instead, it is better to begin with comparisons with the older literature on ethnogenesis as a basis for understanding transnational and diasporic identities in an era of globalization.[20]

In summary, I am arguing for an approach to globalization that examines it as an empirical problem that can be analyzed across societal domains such as politics, economics, ethnic relationships, and civil-society and social movements. The alternative pathways have emerged in the interstices of a world in which people see their communities, democratic institutions, jobs, material culture, and personal relationships being uprooted by distant economic and political institutions that seem unresponsive to their needs. Although the alternative pathways attempt to articulate an alternative to the world of corporate globalization, they are also caught up in it, and their best-laid plans, technologies, knowledges, organizations, and products often go agley when the mainstream political and economic institutions refashion the alternatives. Again, it is the dialectic of opposition and compromise, of incorporation and transformation, that I hope to illuminate.

Methodology

Scholars and activists often have dismissed at least some of the aspects of the alternative pathways as middle class in social address, confused in goals, and timid in repertoires of action. Some scholars may reject the impure mixes of profitability, charity, lifestyle, and other goals along with social-change goals as siphoning off resources from fundamental social-movement struggles based on the goals of protecting or enhancing the rights of workers, women, and ethnic minority groups. Instead, I suggest a more complex approach to alternative pathways that does not involve mere dismissal of the full range of goals. If some people choose to articulate their politics through mixes of scientific research, product innovation, alternative living arrangements, hobbies, charities, consumer purchases, and the creation of new firms, researchers should follow them and understand what is happening. If some types of movements and social-change projects undergo incorporation into mainstream institutions, researchers should understand the patterns and how it is also accompanied by the transformation of activists' goals, knowledge, and design. Such are the questions that can be opened up by the broad comparative analysis that will follow.

Although there is a normative position guiding this book (i.e., that shifts to a more just and sustainable society represent progress), this is not a work of political philosophy, nor does it end with specific policy recommendations. Rather, this book is written in a historical social-science tradition that might be described as a contribution to historical sociology, the comparativist tradition within social anthropology, and the side of science and technology studies concerned with publics and social movements. The social-science method is deductive. In other words, the book contributes to the literature by developing a synthetic conceptual framework that makes possible an understanding of the complex interactions among scientific research fields, technological fields, product design, industrial innovation, social movements, firms, and states. Much as Weber employed his ideal types to explore historical sequences, my point is not to impose a theoretical template on empirical histories, but instead to develop conceptual categories that can serve as guideposts to the study of historical patterns in order to get a better handle on how causality works in historical change.[21]

The nomothetic approach of the comparativist tradition in sociology and anthropology is to use ethnography, history, and primary sources to extract an appropriate level of historical detail for the purposes of shedding light on a selected theoretical problem, in this case the trajectory of environmentally oriented alternative pathways in science and industry over time. Only the theoretically relevant portions of the history and ethnography need to be recounted. History and ethnography operate in the opposite direction; they tend to utilize social and cultural theory as a source of concepts that can help elucidate narratives about historical and cultural particulars. To clarify the method for what is likely to be an interdisciplinary audience, a two-by-two matrix of nomothetic vs. idiographic knowledge crossed with descriptive/analytical vs. normative/prescriptive knowledge may be helpful. (See table I.2.) At a disciplinary level, the ideal types of knowledge are history and ethnography (idiographic and descriptive/analytical); sociology, social anthropology, and political science (nomothetic and descriptive/analytical); policy and legal studies (idiographic and normative/prescriptive); and ethics and philosophy (nomothetic and normative/prescriptive).[22]

A vision of interdisciplinary social science as an integrated conversation informs this book, with the analysis in chapters 4–6 falling into the upper quadrants. The conclusion moves more into the lower quadrants.

Table I.2
Disciplines and types of inquiry.

	Idiographic	Nomothetic
Descriptive/analytical	History Ethnography	Sociology Social anthropology Political science
Normative/prescriptive	Policy studies Legal studies	Ethics Philosophy

The different perspectives can be brought together into a whole as follows: my primary goal is to develop a conceptual framework for understanding alternative pathways for change in scientific, technological, and industrial fields in an era of globalization; I use that framework to conceptualize a comparative historical analysis of a diverse set of alternative pathways and to understand similarities and differences in the late-twentieth-century trends of those pathways; and in the conclusion I use the analysis of the alternative pathways as a basis for a normative inquiry into the justly sustainable society and the kind of societal organization that would facilitate its achievement. By articulating the interdisciplinary frameworks that cross intellectual divisions, including the descriptive/normative divide, I am also articulating a form of critical social science that is opposed to modernist strategies of disciplined and scientistic inquiry that refuse to leave the upper quadrants. I am also operating in a deductive, conventionalist social-science tradition—a tradition that is at odds with the empiricist assumptions of much of the narrowly focused social-science inquiry today, particularly in the Anglophone countries.

Now, it may seem contradictory that a book written in the comparativist tradition of social sciences focuses almost exclusively on the United States. The quick answer to the criticism is that a comparative analysis can be effective when limited to a particular historical time and place, as long as generalizations are not made beyond it. The choice to focus on the United States can also be defended because much of the world views the U.S. as failing to provide global leadership on environmental and justice issues, and in part because of my own limited resources of access to information. The focus on the U.S. creates certain analytical blind spots, many of which I am aware of because of my background in Latin

American studies and my decades of teaching comparative and global studies courses, but it also has the advantage of focusing an analysis of alternative pathways in globalization on what is going on inside the world's superpower. In many cases environmentally oriented innovations are much more developed in other countries, particularly those of northwestern Europe, where the alternative pathways are more highly incorporated into official government and industry circles and less driven by the politics of grassroots opposition. Likewise, in Latin America and other less wealthy regions the alternative pathways take very different forms, such as the focus of environmental movements on reducing poverty and halting development projects that destroy ecosystems and local societies. Such issues of comparison are important but beyond the scope of the book; it is my hope that I will have the time and resources to examine them at some future date.

A second methodological criticism might be raised about the appropriate level of historical detail. In chapters 4–6 I have tried to strike a balance between merely extracting a synopsis of the history and writing the history itself. It would be impossible to write a single book that covered all of the alternative pathways in historical detail, and in the interests of focus and coherence I eliminated the alternative pathways in the health and media fields from the final draft of this book. I have had to grapple with unevenness because in some cases the histories and ethnographies are already written, whereas in others they remain to be written. My goal has been to tell enough of the history to understand the trajectory of organizational, scientific, and technological changes that have occurred since the middle of the twentieth century but not to dwell on all the details. In many cases I have had to assemble the limited history through use of primary sources, and in some cases the process of incorporation and transformation is too recent for much to be said. But rather than drop the discussion where the history is poorly documented or in incipient stages, I have flagged the variation as part of the empirical problem to be examined.

Outline

Chapters 1–3 are more theoretical; they develop a framework for understanding science, technology, and innovation that makes it possible to appreciate and conceptualize systematically the role that environmentally oriented alternative pathways play in industrial innovation in the United States. In an increasingly technological world, scientific research has become the basis of industrial innovation. However, we do not yet have a theory of science that can adequately comprehend the relationship between the goals of activists and advocates in the alternative pathways, who experience gaps in needed expertise, and the scientists and inventors in universities and firms. Chapter 1 provides a theory of scientific change that builds on the legacies of empiricism, conventionalism, and the sociology of scientific knowledge that have come to be known as science studies. I suggest a way of moving beyond constructivism, with its valuable but narrow focus on knowledge claims and microsocial processes, to the broader question of the agendas of research fields and the role of funders in the selection of knowledge that is targeted for development or left undone.

Chapter 2 continues the discussion by developing a theory of science in an era of globalization. Here I address the general historical question of how societies are changing and the specific question of how university-based research is changing. Although I discuss the emerging literature on the increasing ties between science and industrial competitiveness goals, I also explore the countervailing process of "epistemic modernization," or the increased openness of science to lay, public, and civil-society perspectives. Part of the discussion also involves the rise of alternative pathways in science.

Chapter 3 provides a parallel discussion for technology. It develops a critique of simplistic efficiency explanations that is parallel to the critique of simplistic empiricism in science studies, and it builds on the technology studies literature on sociotechnical systems and actor networks to develop the concept of a technological field. The chapter then develops an analysis of technological fields in an era of globalization that is parallel to the discussion in chapter 1 for scientific fields. I introduce the concept of "object conflicts," or definitional struggles over what the technology or product and its appropriate design should be, then outline the alternative pathways associated with industrial innovation.

Chapters 4–6 are more historical in orientation; they develop an analysis of four major types of alternative pathways. Chapters 4 and 5 focus on two types of alternative pathways and their relationship to industrial innovation, empirically grounding the overall argument that social movements and activists have played and are playing a significant role not only in scientific but also industrial innovation, at least in the environmentally oriented cases analyzed here. The industrial opposition movements (IOMs) call for moratoria on unwanted technologies and products, whereas the technology- and product-oriented movements (TPMs) favor the development of alternative technologies and products. In each chapter, I examine the patterns by which governments and mainstream industries incorporate the proposed changes of direction, and how they also transform the original goals, technologies, and products into forms that are complementary to existing ones. Chapters 4 and 5 focus on five industrial fields: food and agriculture, energy, waste and manufacturing, infrastructure, and finance.

Chapter 6 discusses the localization of alternative pathways as a dimension of globalization. Two types of pathways are considered: localist pathways, which advocate the development of economic institutions that are locally controlled and engage in regionally oriented import substitution, and access pathways, which advocate and provide material resources to the least fortunate members of society. Again, I consider the pathways across the five fields and examine the environmental potential of the pathways, but I also examine their limitations.

The focus of localist and access pathways on justice as a primary goal and their wide range of organizational innovations provide the basis for the synthesis in the conclusion, where I use the analysis of the four types of alternative pathways to contribute to the discussions of "just sustainability," and I examine the type of economic organization that could achieve a more justly sustainable society.

1

Retheorizing Scientific Change

In order to develop a conceptual framework for understanding the role of alternative pathways in industrial innovation, it is necessary to begin with existing theories of science, then develop an understanding of how alternative pathways interact with scientific change and industrial innovation. As a social institution, science is enormously important, because it sets the stage of modern politics by circumscribing the horizons of the possible. It proclaims authoritatively what is and can be the case, and it grounds normative projects of public policy and technological innovation in a realism of the possible and impossible. Scientists need not make policy recommendations to be politically influential. By drawing the lines between the true and the untrue, as well as between the possible and the impossible, they can eliminate from consideration normative proposals that are based on assumptions deemed invalid and futures deemed impossible. Although the legitimacy of science as an institution depends on its claim to be relatively innocent from direct influence by political positions, the autonomy of the scientific field is precarious. The scientific field and the doxa that it produces is more like a carefully tended garden than a wilderness preserve. Increasingly, the crucial question for the garden of science in an era of globalization is "Who decides what plants are grown?"[1] To answer this question, it is necessary to begin with a simple but workable definition of "elites." I use the term to refer to networks of people and organizations that control investment decisions and policy making. Though interconnected, they are divided by industrial affiliations and institutional positions, so that political and industrial elites often have sharp internal divisions, and political elites may also clash with industrial elites. Because elites are often divided, they tend not to "breathe together" (literally, to conspire) in a simplistic manner that is evoked by the phrase "ruling class." Often, political and economic elites see their actions as representing the best interests of society, and often they can turn to the elites of the social and natural sciences to support their interpretations of what is possible and impossible.

However, elites are also responsive to pressure from below in the form of consumer preferences, opinion polls, elections, and social-movement action. As a type of elite the leaders of scientific fields are in a delicate position; their research fields undergo profound external influence from economic and political elites in the form of funding preferences for research agendas, but the scientific fields also undergo less profound influence from social movements and general public opinion. Scientists are sensitive to both even as they defend their field's autonomy. The dynamics of scientific fields and their position in society will be the topic of theoretical exploration in this chapter and the next. In order to develop a conceptual framework for addressing the problem adequately—that is, in a way that does not reduce science to ideology but that also escapes from the micropolitics of networks and knowledge construction—it is necessary to revisit the field of science studies and begin with some of its basic arguments.

The Problem of Undone Science

Because political and economic elites possess the resources to water and weed the garden of knowledge, the knowledge tends to grow (to be "selected") in directions that are consistent with the goals of political and economic elites. When social-movement leaders and industry reformers who wish to change our societies look to "Science" for answers to their research questions, they often find an empty space—a special issue of a journal that was never edited, a conference that never took place, an epidemiological study that was never funded—whereas their better-funded adversaries have an arsenal of knowledge to draw on. I call this "the problem of undone science." From the perspective of the activists and reform-oriented innovators, the science that should get done does not get done because there are structures in place that keep it from getting done.[2]

The prioritization of research tends to create huge pockets of undone science that result in the systematic nonexistence of selected fields of research. Where is the university that has all of the following: an electrical engineering department that focuses on distributed and off-grid renewable energy; schools of architecture and urban planning that focus on sustainable design for low-income neighborhoods; a school of agriculture oriented toward sustainable local agriculture; a department of chemistry that works closely with chemically exposed communities and develops green chemistry alternatives; a school of business administration that focuses on developing employee-owned, locally owned, and cooperative businesses; a psychiatry department that explores mind-body therapies as replacements for pharmaceuticals; and a biochemistry department that focuses on food-based neutraceuticals? Although such departments and research clusters exist here and there, and if put together in one place would probably make an interesting and powerful university with unpredictable new synergies, they have not been selected as the dominant research fields and problem areas.

The point need not be overstated. Pierre Bourdieu once noted that the state has a left and a right hand; that is, there are ministries dedicated to issues such as education and welfare and those dedicated to commerce and defense. It is also the case that the university has a left and a right hand. Universities will probably always continue to have a left hand that educates students in citizenship and prepares some for careers of public service. One tends to find such departments and schools among the humanities, social sciences, and professional schools oriented toward the functions of the welfare state, such as schools of public health and social work. But in the current era of globalization, the transition of the research university into an engine of regional development implies that the right hand will tend to become much larger and stronger, and the spaces for developing alternative research fields may become narrower. Even in the left-handed schools and departments within the university, one finds trends toward an orientation to the needs of industry. As a result, the problem of undone science is likely to increase rather than to diminish.[3]

The politics of undone science appear not only in decisions surrounding funding priorities but also in the controversies that envelop the knowledge-making process. Because some methods and equipment cost more than others, the dominant networks tend to have access to the most expensive methods and equipment, and as a result well-funded networks can drown out the alternatives not only through gross productivity but also through access to the preferred methods and the disciplinary institutions that enforce definitions of what is better science. Methods in science are somewhat akin to lawyers in the justice system; the wealthy often have access to the best lawyers, so they have an easier time winning their case. However, knowledge and truth are not infinitely malleable. Because science is fragmented as an institution, it is possible for reformers to go around the consensus of a subfield and recruit advocates and research methods in neighboring fields. In other words, there is the potential for countervailing powers in the scientific field to reduce the dominance of scientific elites. Still, the politics take place in a historical terrain that is increasingly right-handed in the sense of being shaped by the problems, methods, and conceptual frameworks deemed important by industrial and political elites that seek innovative and profitable new products to create jobs, earn foreign exchange, and enhance overall economic and military competitiveness.

Theories of How Knowledge Changes

In order to be of assistance to reformers who use scientific knowledge, social scientists and scientists alike need a theory of science that is not naive with respect both to the epistemic and political authority of science's truth-making machinery and to the often invisible hands that tend the fields of disciplinary knowledge. We need a theory of how knowledge changes in science and of how science is shaped by society, yet we also need a theory that avoids a combination of philosophical relativism and political realism that can reduce scientific knowledge to political ideology. This chapter lays out such a theory by assessing and building on the long interdisciplinary conversation about science that has involved philosophers, historians, and social scientists.

The most basic models of how scientific knowledge changes were developed by philosophers, whose work idealized scientific change in a manner akin to the way that neoclassical economists developed idealized models of markets. Under the empiricist model of science, which in many ways is the "lay" philosophy of working scientists, a sharp division exists between non-observable, theoretical terms and observational terms. Concepts or words that describe unobservables are seen as useful heuristics, or, if one is a philosophical realist, they may present hints of a deep structure of reality that is not yet observable, such as the concepts of viruses and electrons did before machines were developed to transform the theoretical terms into observables. In either case theoretical terms are devices for making generalizations from observations, and science changes when new observations cause scientists to rethink their generalizations. An empiricist may also accept a gradual model of scientific progress based on the subsumption of narrow generalizations or theories by broader ones. If two theories cover the same empirical material, scientists choose between them by finding a point where they predict different observations, then the scientists design a crucial experiment to determine which prediction is more accurate. If the theories are evidentially indistinguishable, a true empiricist will bite the bullet and say there is no ground for choosing between the theories.[4]

Alternative views usually begin with the recognition that science can have other rational grounds for distinguishing among theories, such as consistency with other theories and a combination of internal consistency and simplicity. By connecting theories to each other as much as to observations, the ground is set for recognizing the theory-ladenness of observations and methods. The long tradition of conventionalism, which dates back to the early-twentieth-century French scientists Henri Poincaré and Pierre Duhem, argues that methods and observations are only interpretable within a theoretical system. Furthermore, because theories can be adjusted to new data, it is not easy to design a crucial experiment or decide upon a crucial observation that would allow a clear choice between two broad theoretical systems. In the face of what appears to be contradictory evidence, a defender of an existing theoretical system can make various moves: argue that the methods behind the new empirical evidence are flawed, claim that the interpretation of the data is wrong, or modify a subtheory without jettisoning the broader theoretical system.[5]

A conventionalist approach to knowledge change has the advantage of bringing the model closer to scientific practice and recognizing that new empirical observations do not easily resolve major theoretical controversies, especially when large networks of scientists have substantial intellectual and material investments in existing theoretical systems and associated research programs. Instead, the history of science often narrates stories of scientists who line up behind one theoretical and method-ological system in opposition to another group of scientists. Resolution of the controversy is grounded in logical argumentation and displays of evidence, but it also requires negotiation between the sides over what counts as evidence and what methodologies and research designs are considered acceptable. The accumulation of evidence gradually tends to put one side increasingly on the defensive and leads some of the advo-cates to be persuaded by the opposing view. As Max Planck observed and Thomas Kuhn popularized, sometimes one side must retire or die before the controversy is fully resolved. Moreover, if one side has better access to laboratories and other research resources, it is in a stronger position. However, just as having the funds to hire the best lawyers is helpful in a court trial but does not guarantee the outcome, a network that controls the best laboratories is not guaranteed of winning a scien-tific controversy.[6]

Perhaps the best-known conventionalist account of science is Thomas Kuhn's book *The Structure of Scientific Revolutions*. In Kuhn's model, scientists in a research field labor under a paradigm, or a specialist research culture of observations, theories, exemplars, methods, and people. The theories of a paradigm do not undergo revision as the result of a single crucial experiment; rather, the paradigm undergoes a slow and steady erosion by the accumulation of anomalies. As new research results and problems come to shake faith in the existing paradigm, an alterna-tive is proposed, and the outcome is resolved according to the conventionalist model of mixes of empirical evidence, consistency argu-ments, and negotiation of methods and results. When the challenging perspective is triumphant, there is a scientific revolution, and scientists who labor under the new paradigm settle in to work under the condi-tions of normal science.[7]

The Structure of Scientific Revolutions currently has more than 7,000 citations in the various science citation indices, and some have claimed that it is the most-cited book written in the twentieth century. The Kuhnian model of scientific change was widely influential, and some young scientists read it with the aspiration of becoming the leader of the next revolution rather than a mere cog in the machine of normal science. Social scientists who work on social movements, environmental issues, and other fields that are not directly connected with science and technology studies also tend to talk about science in terms of paradigms and revolutions. However, after half a century a substantial literature has developed to reveal the book's shortcomings. Perhaps the greatest shortcoming is the idea that a specific scientific field is governed by a monolithic paradigm, and likewise the converse idea that some scientific fields, especially the social sciences, are pre-paradigmatic anarchies of immature science. Instead, attention has come to focus largely on the diversity of the theoretical and methodological differences among networks of scientists within a field. Although there are many versions of what has become known as the sociology of scientific knowledge, most of them recognize the intense power of networks and their competition for recognition in science. Scientists within a research field do not all march to the tune of the same paradigmatic piper; rather, research fields are often characterized by disagreement and controversy over empirical claims, proper methods, and conceptual categories.[8]

To formalize the alternative, agonistic view of research fields, a research field is characterized by relations of cooperation and conflict among advocates of different conceptual frameworks, research methods, and problem areas. An individual scientist usually is engaged in more than one research program, that is, a bundle of research that brings together a method and theoretical framework to bear on a research topic. Scientists who work on similar research programs can be said to consist of a specialty network; they are the most competent to review each other's work, and they form what Harry Collins has called the "core set" of actors when a controversy erupts. However, other scientists may have some overlap in expertise with a portion of the bundling of theory, method, and problem area that occurs in a specialty network. As a result, peer review is also possible by neighboring colleagues who hold expertise in a portion of an individual's research program but not the entire program, such as the methods or conceptual frameworks but not the problem area. One can also determine the degree of proximity and therefore to some extent the value of the peer review by assessing how much of a theoretical framework, set of methods, and problem area is shared.[9]

When controversies erupt within networks of proximate colleagues, peer review by non-proximate but neighboring colleagues provides for a system of checks and balances in science, and it provides editors, funders, and other gatekeepers with a mechanism for sorting out disagreements. If a challenger scientist has developed a new method or generalization, or if the scientist has produced a new empirical finding that is not recognized by the specialty network, the challenger scientist has recourse to appeal by attempting to persuade the non-proximate colleagues who share a research culture along one or more of the dimensions. The process is both intellectual and social at the same time; in order to make knowledge change, scientists must mobilize both convincing arguments and convinced colleagues.

This view of scientific change is more or less the way researchers in the field of science and technology studies approach knowledge today. Most empiricists have ceded at least some ground to conventionalist arguments, and most social scientists with a conventionalist orientation have recognized that although knowledge making involves socially negotiated argumentation, scientists must still convince their colleagues with evidence and logical argument, and consequently the epistemological status of scientific knowledge is not equivalent to political ideology. One can take science off its pedestal of a naive empiricism without giving up the claim that scientific knowledge is, like other forms of occupational expertise, generally superior to that of non-experts, at least on topics where the expertise is well developed, empirically grounded, and openly vetted.

In this sense we can say that scientific knowledge is socially constructed. The term is used here not as a philosophical claim but instead as a limited empirical generalization from social-science research, which has found that scientists must vet differences over theories, methods, and the interpretation of observations through a social institution that relies on negotiated assessments among more or less proximate peers. As a result scientists generally must make complex judgments, much like a jury that is evaluating various arguments from the defense and prosecution, rather than resolve disputes through the simplicity of the single, definitive, crucial experiment. The process of understanding how the world works in science is similar to the processes in other rationalized modern professions, such as lawyers establishing the facts of a case, doctors conferring over the diagnosis of a disease, or mechanics assessing different explanations of what is wrong with the car. Although scientific knowledge, like other forms of expert knowledge, is fallible, I will assume that scientific knowledge about the natural world, like social scientific knowledge about the social world, is generally better than that of non-experts—although not always so, as literatures on lay and non-Western knowledge demonstrate—and that the knowledge accumulates, even if it sometimes undergoes theoretical reconceptualization.[10]

Beyond the Empiricist-Conventionalist Debate

Although the philosophically oriented models of scientific change are valuable as a starting point, they are in a sense devoid of content. In other words, it hardly matters if one is talking about physics or psychology; the philosophical and sociological models are concerned with general processes by which a community of experts determines that one theory or observation (or theory-observation-method bundle) is superior to another. The problem is the construction of knowledge, of how knowledge is built up from the basis of previous research and vetted for general acceptance within an expert community of scientists. In this book I am more concerned with the content of scientific fields, that is, the question of which knowledge comes to be selected as deserving attention and which knowledge is considered not worth pursuing. I term this second problem the "selection" of knowledge, in contrast with the construction of knowledge. The term may invite some confusion, because of the widespread association of selection with evolutionary theory. However, one might remember that Charles Darwin began his discussion of natural selection with the human selection of domesticated plants. The concept of selection is understood here in the root etymological sense of choosing (as in the selection of candidates, theories, or products), which can include ranges of choice that are imposed on the less powerful. In my view, the primary question for science and technology studies in an era of globalization is no longer the constructivist question of how scientific knowledge is socially negotiated or shaped, but instead the structural question of what science is selected to be done. To begin to answer that question, we must first turn to the reward system in science.[11]

In the early philosophical and sociological models of scientific fields, research takes place inside a bubble, as if it were socially isolated from society. From that perspective, research fields grow and develop organically from new questions that emerge from new answers. The agenda of research topics flows from basic curiosity about a piece of the world, but the agenda is not influenced by societal demands on science as an institution. This view, based on an assumption of a high level of autonomy for the scientific field, represents a simplified or idealized model of scientific change.

The autonomy assumption represented not only the peace in the feud among empiricist and conventionalist theories of how scientific knowledge changes, but also between the philosophical accounts and the modernist sociological accounts of the reward system. A reward system under conditions of autonomy allocates prestige to scientists based on their ability to solve problems in a field. Robert Merton is generally given credit for having first described the reward system, but Warren Hagstrom's subsequent work drew out two crucial points. First, the reward system has some features similar to pre-capitalist gift exchange. Scientists in the ideal world of pure science give research away, but they do so because they receive recognition in the form of citations, prizes, or general renown among peers. Even though scientists have given away their research, they remain attached to it because it is the source of prestige. As a result, they will defend their work against attack to preserve its scientific aura in the peer networks. Second, as in a gift exchange system scientists also compete for recognition; the very people on whom one is dependent for recognition are also competing for the same recognition. It is as if the athletes competing for an Olympic gold medal also made up the panel of judges. Under such circumstances backstabbing and strategic alliances can be expected, not because the stakes are so low, as is commonly said of academia, but because the system is set up to hold in constant tension relations of cooperation and competition. The primary check on duplicity is provided by the existence of non-proximate and multiple peer review.[12]

In the 1970s two significant theoretical developments began to open up the analysis of the reward system from the autonomy assumption to a framework that is compatible with the study of science and globalization. Bourdieu's convertibility thesis focused on how scientists transform economic resources into the symbolic capital of publications and prestige, and vice versa. He also argued that such conversion strategies were crucial in agonistic struggles over credit. The convertibility thesis became increasingly valuable as historians and ethnographers began to pay more attention to the problem of how funding flows affected research agendas. The other development was Bruno Latour and Steve Woolgar's analysis of how credibility is a crucial factor in the investment cycles of material and symbolic capital. They argued that successive publications give scientists the credibility to gain access to funding that leads to increased research and more publications, which in turn leads to more funding, and so on. The accumulation of reputation or credibility is concretized in the curriculum vitae and can be traded as collateral for material resources for future research. In other words, as the reputation or track record of a scientist increases, it generally becomes easier to obtain larger grants, better positions, higher salaries, lower teaching loads, better graduate students, and better physical research space and equipment. Provided that access to increased inputs results in continuing recognition for the scientist's research, there is what they termed a "cycle of credibility."[13]

The increasing-returns dynamic of the Latour-Woolgar model was anticipated somewhat by Merton's cumulative advantage theory of scientific careers—the idea that in science the rich (in prestige) get richer and the poor get huge teaching loads—but the Latour-Woolgar model suggests that increasing returns would operate regardless of the starting point. Their model also showed how the success of a scientist's publications is related to demand: successful research is cited research, that is, research that is used by other scientists to build their own research. Consequently, as one's own research is embedded in other scientists' research programs, and, as the number of users increases, the cited scientist is increasingly able to withstand challenges. In what amounts to an early version of actor-network theory, the higher the number of positive citations, the more widely accepted a piece of research is, and the more it approaches consensus knowledge. It is not necessary to enter into the question of whether the high use of a scientist's research is caused by the scientist's successful marketing, strong social networks, and impeccable pedigree, or because the research happens to represent a portion of the world in an accurate and novel way. The answer is simply that a mix of the two occurs, and the factors operate synergistically.[14]

The work of Bourdieu and Latour and Woolgar helped to undermine the autonomy assumption by opening up the reward system to include material capital and by showing how in a sense the goal of the scientific game is to accumulate a high mix of both symbolic capital (citations, prizes, prestigious appointments, and successful students) and material capital (grants, laboratory space, equipment, postdocs, and graduate student research assistants). The opening to material capital punctures the bubble of autonomy and shows how knowledge making is contingent on external sources of funding, with the exception of a few fields that have negligible research costs. In other words, funding shapes what science can and will be done as well as what remains undone. However, one must be careful with this argument, because it can turn into a simplistic form of externalist, economic determinism. To further develop a post-autonomist theory of scientific change as knowledge selection, this chapter will analyze how individual scientists select research programs as well as the more general issue of the rise and fall of research fields. Although the discussion may seem technical, it is fundamental for understanding the problem of undone science and the potential for activism and social movements to play a role in knowledge generation and in technological innovation.

The Selection of Research Programs

It is possible to find research fields in which there is little or no controversy, but it is more typical for the research front of a research field to be characterized by controversies over methods, conceptual frameworks, and assessments of what constitutes an important and trivial problem area. However, the existence of controversies does not necessarily imply a level playing field of pluralistic research networks; hence, Kuhn's paradigm concept does flag a condition that is often seen. It is often the case that research fields have one or more dominant networks that control the lion's share of resources in the field. The dominant network is the in-group that has hegemony over the major graduate departments, journals, professional society positions, and grant-awarding institutions. But the position of the dominant network is not stable; it is constantly facing challenges by new networks of researchers who are importing new methods or concepts for a problem, or who are attempting to divert resources for a method or conceptual framework to new problem areas. Although it is possible to find suggestions of a paradigm in the sense of consensus knowledge, it is also the case that at the research front most research fields are characterized by controversy.[15]

Lack of consensus is not a sign of an immature science or a paradigm in the throes of revolution but instead an indication of the vigorous vetting process that occurs through complex linkages of cooperation and competition. Controversy over what the important problems should be, what the best methods are, and how the problems should be defined are all aspects of "normal" science, not, as Kuhn would argue, symptoms of the pre-paradigmatic phase of a research field's history. As a controversy is resolved, the change may not be as simplistic and dramatic as a paradigm change; instead, the challenger networks of theory-method-problem bundles may find themselves incorporated into the dominant networks but also transformed by the incorporation process, often through theoretical and/or methodological syntheses.

In view of the network structure of scientific fields, the process of choosing a research program is both an intellectual and political investment. Consider a scientist who, during the wind-down stage of a significant project or merely during a pause in a busy schedule, steps back to assess what research projects should come next. A wide variety of factors influence the decision, as has been noted as long ago as the work of Hagstrom. The scientist will probably make an assessment of the risk/reward ratios for the new research project, somewhat akin to an investor who is selecting a stock. For example, there may be a hot new growth field that has a lot of buzz, many new entrants, and upward funding curves. As Henry Menard noted, if a scientist makes a significant contribution to a new field, the rewards will be great, and consequently new fields pose great opportunities to young risk takers. However, the field may not pan out, the funding may dry up, and a bandwagon of new entrants may quickly crowd the field. For more established scientists the costs of retooling and the high risks of failure in a new field may make entrance less attractive. In contrast, by returning to an ongoing research program there are lower entry costs (both intellectual and material), and although the returns may not be as great, they are less risky. Still, the "bond" option of scientific investing poses risks of its own; scientists who pursue that option face diminishing returns of a method to the same research topic. Scientists with larger pools of accumulated capital may consequently diversify their research program investments by putting some of their resources into established research programs and some into high-risk, high-return programs, and they can also have the resources to migrate more successfully into high-risk, high-return fields if their original investments pan out.[16]

One of the risks of new fields, especially for junior researchers who are often the first colonizers, is the well-known pattern, first characterized by Merton, whereby less well-known scientists who may have priority will tend to receive less credit than better-known scientists who do not have priority. One can also generalize the argument, as Margaret Rossiter has done, to notice that credit flows not only to more senior scientists but disproportionately to privileged social categories, such as men over women. Young scientists in a new, high-risk field may at first welcome the entry of more senior scientists into the field because they legitimate the field and help open up the doors to greater funding and access to publications, but the entry of the silverbacks can be a mixed blessing for the young founders, because the silverbacks will tend to redefine the field, absorb future credit, and reapportion credit to themselves and their own students over the founders, especially if the founders lack the initial network linkages of the high-prestige late entrants. There can even be a rubbing out of the achievements of the lesser known researchers as credit is absorbed by the better-known researchers.[17]

Two points should be made about the economic model of the decision-making process of the scientist. First, the model can be accommodated to either an autonomist or non-autonomist view of science. In other words, the existence of high-risk, high-reward fields can be an outcome of an internal logic of problem solving, and scientists can capture pools of externally controlled funding for the new field by convincing funders to alter their priorities. However, the general priorities of funding patterns, which in turn are a significant factor in how hot or cold a research field is, are often determined by extra-scientific actors, and the pattern is increasingly away from control of broad agenda setting by scientists, if they ever had such control. Historians of science and funding have documented the effects of extramural funding on research priorities in a variety of fields, from physics, chemistry, and engineering to biology, medicine, and the social sciences. From the viewpoint of the scientist on the ground, the system may appear to exist in an autonomous bubble that develops according to its own internal logic, but when one steps back and follows the money, it is clear that funding flows help some fields to prosper while others wither on the vine.[18]

Second, the economic metaphor of the scientist as a rational actor who assesses risks and benefits before making investments of research effort should be seen as providing a helpful but ultimately incomplete picture. There are many ways in which investment decisions are not rational in the sense of decisions that are based on calculations to optimize recognition and other rewards. One example is the scientist who clings to an old research program, even when the evidence has been overwhelmingly in favor of divestment for a long period of time. As in the financial world there can be unsuccessful and incompetent investors in the scientific world. Another example is the scientist who invests in a research program because of a sense of its societal value. The example is particularly important for developing a structural understanding of the career risks faced by scientists who are aligned with social movements and with social-change values; those scientists may choose to invest in research programs that result in lower prestige and marginalization, if not outright suppression. They are not bad investors as much as people who understand the tradeoff between socially responsible scientific research and career advancement. Scientists who pursue out-of-favor research that is linked to social movements and social responsibility values sometimes attempt to make an alignment between the societal value and funding availability. As Adele Clarke has demonstrated, scientists who work with social movements may also negotiate a "quid pro quo" in which movement and advocacy organizations exchange financial support for a research program that is not what they originally wanted but is closer to the scientist's needs for research that will be valued within the scientific field. A third example of decision making that is not intuitively accommodated to a prestige-optimizing model is the scientist who opts for applied research projects that have high financial rewards, such as licensing opportunities for patents, but much lower status rewards, such as prestigious publications. In the extreme case the scientist is transformed into an inventor and exits from the prestige game of the scientific field. The transformation, which provides a linkage between the scientific and industrial field, can take diverse forms. For example, the scientist may pursue technological research that is funded by a large corporation, or the scientist may engage in technology development in an entrepreneurial setting. Cross-cutting the settings can be a mix of values from pure profit seeking to social and environmental amelioration.[19]

Given the shortcomings of a prestige-optimizing model of scientific decision making, one might argue that choosing a research program could be likened less to an investment decision and more to a decision to move into a new neighborhood or to look for a new job. By shifting research programs, a scientist also shifts the reference specialty network and faces the inevitable adjustments of getting to know new neighbors and colleagues. Depending on how far one moves into the proximate peer networks, one can face increasing tests to prove oneself. In the new specialty network a scientist's accumulated reputation may not matter as much, and one must rebuild relationships with new colleagues. As with any move, the process can entail a range of experiences from liberation from old rivalries and antagonisms to the travails of intellectual hazing from anonymous peer reviewers and citation taxes levied by journal editors who want a specific network cited before accepting a paper for publication.

The Selection of Research Fields

So far my analysis has focused on the issues a scientist may face as an individual when selecting research programs and assessing the option to do research in areas of undone science that is relevant to alternative pathways. However, the decisions of individuals take place within broader historical changes by which whole research fields come into and go out of favor. As funding priorities set by governments, foundations, and private corporations shift, researchers will tend to follow the money at an aggregate level, even if some individuals select problem areas that go against the incentive structure. For example, a problem area with growing government and industrial research funding, substantial startup opportunities in the private sector, departments in the top universities, and high salaries (i.e., nanotechnology research in the United States at the time of writing) will attract many researchers. Because money spent on one problem area of a research field is money not spent on another, the size and shape of problem areas follow the funding flows, and the scientists who come to occupy a backwater will find themselves and their students excluded from the rewards of prizes, large grants, and the circulation of senior hires among elite departments. Where individual careers are evaluated in terms of dollar value of sponsored research at the tenure point, the pressures to move into the mainstream fields will be even stronger. In other words, the aggregate funding flows that shape the status hierarchies within and among research fields exert a downward pressure on the portfolio decisions of individual researchers, just as the decisions of individual scientists about what should be the next problem area for a research front exert upward pressure on the funders' decisions to reallocate portfolios. The availability of funding makes new research possible, but the positive results of the new research encourage funders to increase support. Funding will tend to continue to grow as long as the momentum of positive results continues and as long as political and economic elites target the fields for further development.[20]

Although there is significant historical variation in the growth and decline of specific research fields, there is also some value in developing a general model of the economic dimensions of the cycle. In this section I will develop a general economic model of the growth dynamics of research fields, then relate the model to the problem of undone science. To begin, if the funding rate for a new research field does not keep pace with the growth of scientists who have decided to shift into the field, then in the short term competition for funding and other resources will increase. Likewise, in neighboring research fields that are more established and not undergoing the excitement of a dramatic new breakthrough and/or an increase in funding, scientists will begin to migrate out to join the new growth field, and (other things being equal) competition within the old, declining research field will decline. In the short term, an equilibrium mechanism across the fields tends to adjust the expectations of higher recognition with the realities of higher competition, so that individuals face a tradeoff between high-risk, high-reward fields and low-risk, low-reward fields.

If funders agree that the new research field is worth supporting, they will reallocate their portfolios toward the new research field, and (in the absence of overall increases in funding) they will cut the research in the older neighboring fields. Their action will tend to equilibrate the costs of entry across the research fields over the longer term. The declining research field will now be doubly unattractive: not only will the promise of recognition be lower as the field size decreases, but the competition for funding relative to the new field will have increased due to the cuts in research funding. Likewise, the new field still has the higher promise of recognition, but (again, other things being equal) the increases in funding mean that competitive rate for funding has gone down relative to the old field (even if not in absolute terms). As a result under these conditions, the reallocation of funding accelerates the growth in the new field and the decline in the old field.

Just as a small town becomes a large city, when a research field grows the original research network will mushroom so that it is no longer a small-scale invisible college that met Derek de Solla Price's criterion of everyone knowing everyone else. At the beginning a research field might be somewhat interdisciplinary, with scientists from diverse subfields or disciplines reading each other's work. As the research field grows in size, each of the areas represented by a few scientists, or even one research group, might become occupied by a whole specialty network of research groups. New research groups will enter and bring to the research field slightly different problems, methods, and conceptual frameworks. In short, as the field grows in size, it will become more diverse. New conceptual and methodological frameworks from different disciplinary specialties will be drawn into the problem area. A field with a few competing specialty networks may find itself now having several competing networks as specialists in new research fields enter the field. Old-timers may lament the good old days when conferences were small and everyone knew everyone else, but they may also welcome the growth in prestige and opportunity that the success of the field has brought to them and their students.[21]

However, the promising new field does not remain in boom mode forever. There may be a long-run shift from sponsorship by government and foundation sources to industrial sources. With the change in funding, both research problems and intellectual property regimes shift, and some scientists may find other, more basic research fields more attractive for their reputational goals. Even in the absence of the shift toward technology and product innovation, researchers in the same field will tend to experience diminishing returns unless new methods can be brought to bear on the problems. If a field has already grown to a large extent and become fairly diverse, then (other things being equal) the probability of new methods leading to new breakthroughs will be lower. The field will begin to settle into a steady state (a term I prefer to "normal science") of incremental advances in knowledge. In turn, the field may undergo diminishing rates of citation in journals with lower readerships for each new study. Scientists and especially their students will begin to leave the field as the possibilities of new breakthroughs and status enhancement emerge elsewhere. Eventually, funding will also slow down as new boom cycles emerge elsewhere and the once growing field now becomes an established or even backwater field with declining rates of citation and funding.[22]

It is not necessary to assume that the cycle of growth and decline outlined here applies to all research fields. In mathematics and in some of the social sciences, research costs are low and researchers may follow their personal interests or those of small networks of colleagues. When they are free to follow their own interests, they are likely to spread out into an increasingly diverse number of problem areas. As the size of specialty networks declines, theoretical and conceptual issues lose salience because there is no one to provide alternative views on the same empirical problem. Specialty networks per se cease to exist; there are only more proximate and distant neighbors who share some knowledge of each other's methods, conceptual systems, and topical problem areas. Under such conditions, they read each other's work out of a need to maintain institutional support rather than a sense of building together a collective understanding of a common problem area. This anomic type of research field could, in theory, exist in any field that has low research costs.[23]

One solution to scientific anomie is involution: a small number of researchers read each other carefully and cite each other's work, but the citation pattern is more of a closed cluster than an open network. Their work tends not to be cited outside the narrow research field, and the members of a citation ring may develop special languages and methods that make their work especially difficult for members of neighboring research fields to find their work accessible. Rather than fading off into the overlapping networks of neighboring research fields, in which research in one field has spillover effects on another field, the involuted field has closed in on itself. Sealed off and hermetic, the field will tend to be less likely to produce research of value to activists.

Yet scientific involution is probably less responsible for producing undone science than the general trend for government funding to emphasize research fields that advance national competitiveness and for a growing research field to shift toward applied science and technology.

When a mature research field shifts to an emphasis on patenting and licensing, funding will tend to diversify out from public sources, such as federal government research grants, to industrial support. More generally, when a research field undergoes growth, it requires new resources to fuel the growth, and it will tend to seek funding resources from industry. When industrial funding sources increase, opportunities for alignment with general societal benefit and projects that are coherent with social-movement goals will be reduced, or they will have to be recast in ways that are made compatible with industrial goals.[24]

However, because government research is increasingly tied to industrial innovation goals, even in the absence of increasing amounts of direct support from industry the overall funding portfolio for a research field may be oriented toward technological innovation and industrial competitiveness goals from the outset. Researchers who opt to do work in conflict with the competitiveness goals set by economic and political elites, such as on the environmental and health risks of new technologies, may find limited funding budgets with correspondingly diminished field sizes, citation rates, and opportunities for career advancement. They may still opt to pursue research programs that are aligned with social-change goals, and they may do so by cobbling together limited available funding sources, including their own resources, or by dedicating a portion of their research portfolio to pro bono research. In doing so they are swimming against the tide and creating conditions for their own marginalization. Rather than dedicating all resources toward achieving prestige as a founder of a large and growing research field, to the extent that they dedicate their research resources toward undone science, they have opted for a career path of anonymity. They risk becoming a flounder in a field that may be perceived to be tainted by political interest or merely dismissed as boring because it is not on the cutting edge.

Conclusions

In this chapter I have outlined a theory of scientific change and innovation that is appropriate for an era of globalization. Although the history of science can produce some cases that approximate Kuhnian paradigm revolutions, I have suggested an alternative approach for understanding change and innovation in science. Certainly there are consensus shifts in science, but Kuhn's focus on sudden and dramatic changes in the form of revolutions diverts attention from the deeper and more pervasive dynamic of scientific change: the ongoing differentiation and diversification of scientific fields as new research fields are both created by and the creators of funding sources. Rather than see science as moving in a circle of revolution, normal science, and back to revolution, I draw attention to the deeper pattern of innovation in science that occurs with the differentiation of new research fields.[25]

In contrast with Kuhn's pairing of the scientific revolutionary and the inglorious normal scientist, the drone who does the mop-up work of the revolutionaries and is consigned to the dustbin of history, the framework outlined here suggests an alternative pairing. The founder of a new field recognizes opportunities for innovation, especially when the opportunities are not yet obvious to peers, and the founder colonizes new areas of knowledge when the risks are high. The founder does not cause the growth of the field, but neither does the shift in external funding priorities cause the growth of the field. Rather, there is an interactive process in which the founder is an actor in the structured structuring of the history of the growth and decline of a research field. Although the founders of a research field may eventually lose credit during the secondary succession that involves colonization by the dominant networks of adjacent fields, the founders who do not suffer from obliteration through incorporation will tend to enjoy great prestige in science.[26]

In contrast with the figure of the founder, the flounder chooses to stick with research programs associated with research fields where funding, citations, and number of researchers are stagnant, small, or even declining. The flounder may have good reasons for doing so, including an alignment with a social movement's political goals. Flounders will tend to work in the basements of the funding systems (and often of the research buildings) on topics that their peers and funders consider to be unimportant or to have backwater status. They may have graduated from universities that lack the funding to attract the best students, and they may be faculty members who are condemned to social reproduction by watching with chagrin as even their stellar students achieve positions in second-tier institutions. Flounders may even find that they are alone, or located in an anomic field where peer interest in one's research approaches zero. After recognition of their position sets in at mid career, they may opt to leave the game of science to develop applied research for clients, who may range from private-sector firms to social movements.

Although research conducted by scientists at the behest of social movements (or even research programs that are not aligned with the goals of dominant industries) may cast the scientist in the position of a flounder, it is not the only outcome. Clever scientists may be able to produce an alignment of social-movement goals with goals associated with large, growing research fields and even large industrial corporations. For the intellectually, politically, and strategically brilliant, there is a way out of the dilemma, especially if they are located in a prestigious institution and can leverage the institution's halo effect to put a new topic on the intellectual map. If the scientist is able to solve the dilemma, the research may garner for the scientist both high social prestige outside the scientific field and high recognition within it. Such possibilities exist, and their frequency and conditions for success can be studied empirically. The point here is not to make an empirical claim—that all scientists who choose research programs associated with social-movement goals will become flounders instead of founders, or even that there is a valence for such work to produce flounder status—but instead to outline a theory of the dynamics of a system in which such decisions take place, career risks are experienced, and issues can be conceptualized and studied. Because the system is set up so that certain areas of science will be well tended while others will be left to wither on the vine, the scientific field will develop historically to have large areas of undone science. The pockets of undone science will tend to include knowledge that would be especially valuable to the building of alternative pathways.

2

Science in an Era of Globalization

By removing the autonomy assumption as described in the previous chapter, it becomes easier to understand the political and economic dimensions of science. The rise and the fall of research fields are not merely products of the unfolding results of theoretical and empirical inquiry in a quasi-autonomous space; the histories of research fields are also shaped by the agenda-setting goals of funders and social movements. The perspective outlined in the preceding chapter adds content to the philosophical accounts of how knowledge changes by providing a means for examining why some types of knowledge flourish while others wither on the vine.

However, the framework still suffers from a fundamental limitation. The framework is temporal and dynamic in the sense of providing a model of how science changes, but it is not yet historical. There is no theorization of how science changes differently in different historical periods. The framework of the previous chapter could be applied to scientific fields at any point in their history, as long as there is a reward system based on peer review and a need for external funding. To remedy the shortcoming, this chapter will develop a broader historical perspective on science in an era of globalization and the emergence of alternative pathways in science.

The Alignment of Agendas

Scientific knowledge in an era of globalization is both increasingly esoteric (that is, characterized by increasingly technical methods, differentiated roles, and complex equipment) and increasingly exoteric (in the sense of being subject to scrutiny by governments, the private sector, and civil society). Non-scientists need scientific expertise to accomplish their goals of economic and social provisioning in an increasingly competitive world, but they also want to direct an expertise that is increasingly difficult to understand. One way to comprehend both the esoteric and the exoteric dimensions of change in science is through an analysis of the politics of alignment.

For centuries, there has been an alignment between scientific knowledge and the practical needs of industry. The vacuum of the air pump, the problem upon which the edifice of the modern experiment and Boyle's Law were developed, was not merely an interesting topic of disinterested speculation; it was also a practical object of interest to mining firms that needed to pump water out of their mines. Long ago, Boris Hessen showed how the industrial problem of the steam engine enabled the transition to thermodynamics in physics, and likewise he showed how much of modern science was linked to military and industrial applications. There were also ideological benefits of early modern physics, such as the alignment that early English scientists produced between their atomistic views of an inert Nature in contrast with the pantheism of the radical political movements such as the Levellers. By the twentieth century, military and industrial funding shaped the research agendas of whole scientific disciplines. The alignment of research agendas occurred not only for applied fields such as engineering and medicine but also for the social sciences and even the so-called pure sciences, such as physics.[1]

The term "alignment" draws attention to the agency of scientists in shaping their shapers. The goals of the patrons of science can be ill-defined, and because political and economic elites do not possess technical expertise and are divided among themselves, they can, to a certain degree, be told what they want. As a result, the alignment of the interests of elites and those of the scientists is an ever-changing process of negotiation. The elites who control the purse strings of funding are not merely unmoved movers; their goals also shift. For example, with the winding down of the Cold War and the rise of neoliberal globalism, state-based funding shifted toward economic competitiveness agendas. In turn, with the rise of terrorism, competitiveness goals have been grafted onto a security state with new scientific and technological needs.[2]

A substantial social-science literature has developed on the topic of alignment in the context of transformations of scientific research during the era of globalization. Whether one calls the general historical change "mode two knowledge production," "academic capitalism," "the enterprise university," "audit culture," "the triple helix," "degrees of compromise," or "impure cultures," there is considerable documentation of the ways in which state and industry are intervening to establish research priorities for the scientific field. This section will focus on some of the administrative changes that are occurring within the university. The choice is based on the assumption of the importance of the university to the scientific field; however, similar processes occur in other institutions, such as the corporate research and development laboratory. As Roli Varma has documented, in corporate research and development centers management has increasingly pushed scientists to generate research that is more closely aligned with immediate business needs than with long-term technical knowledge. Funding has also shifted from general corporate sources to business divisions, which have reoriented research priorities from basic research toward development.[3]

In the university there is a general increase in administrative control over workers, including scientists as researchers and professors. The trends are well known: the differentiation of academic positions to include non-tenure-line appointments based on teaching or research, the increasing levels of reporting and monitoring of faculty research, the intensified links between research funding and salary adjustments, the development of technology transfer offices and stricter monitoring of faculty research for patent potential, the encouragement of corporate funding of research centers and laboratories, and the replacement of traditional departments with research centers. Administrators establish strategic goals, which are generally aligned with industrial development priorities, then enforce their goals on academic departments through budgets. They utilize a combination of zero-based budgeting, which puts all programs under scrutiny for possible elimination and targets for growth only the programs that contribute to strategic goals, and revenue-based budgeting, which allocates new resources based on algorithms of teaching and research revenue. Through the politics of budgets, academic departments can be brought into alignment with strategic goals. Database management programs can even push the process of monitoring down to the level of the revenue streams generated by individual faculty members.[4]

The alterations in the local reward systems represent a mechanism through which globalization affects the portfolio decisions of departments and individuals, and, through their decisions, the contours of done and undone science across research fields. Where there is a federal government administration that has directly suppressed and opposed some scientific fields, such as research on global warming during the presidency of George W. Bush, such fields will likely not become priorities in a university's strategic plan. Instead, the focus of investment and administrative attention will tend to be on the "T's" of corporate and political interest: IT (information technology), BT (biotechnology), and NT (nanotechnology).[5]

Even research fields that have low research costs are not immune from pressures to align with strategic goals. For example, the social sciences and the humanities tend to have lower demands to produce external funding. However, the reduced demands for research funding are usually accompanied by a greater emphasis on teaching, and pressures on research can develop through teaching programs and variations in student demand. Revenue-based budgeting will create pressures to alter academic degree programs, and with them hiring priorities and research agendas, toward areas of high student demand. For example, during the last 30 years of the twentieth century demand for career-oriented undergraduate degrees in the United States increased between fivefold and tenfold, whereas demand declined or rose only marginally for the traditional liberal arts fields. As I have seen as a former department chair in my own university, the changed demand structure creates incentives for humanities and social science departments to partner with high-demand majors (such as information technology, management, or engineering), and consequently to begin to shift their teaching, their personnel, and ultimately their research portfolios toward fields that are directly linked to student demand, trends in job growth, and industrial priorities. Likewise, where multi-disciplinary grants are highly valued, a social scientist or a humanist will tend to be placed in a position of handling the ELSI (ethical, legal, and social implications) dimensions of the project. One can embrace the change as the new liberal arts or lament it as the death of the old liberal arts, but both perspectives recognize that a transformation is taking place.[6]

The conditions of alignment with industrial priorities make it increasingly difficult for researchers to wrap themselves in the flag of academic freedom, to pursue research programs that are in conflict with strategic goals and aligned with social-movement goals, and, in short, to operate outside the mainstream of their fields and departments. But such work does proceed. As I shall now argue, the changes in scientific research in an era of globalization cannot be reduced to the increasing hegemony of industrial research priorities over scientific research. A countervailing process has also emerged.

Epistemic Modernization

The globalized academy with its new research and development centers constitutes one way in which the degrees of freedom for scientists are constrained from the outside, but it by no means represents the whole story. Just as there is increasing scrutiny over the selection of research programs from above, there is also increasing scrutiny of science from below. I call this second change the "epistemic modernization" of science, a term that draws on the concepts of "reflexive modernization" and "ecological modernization." Such general concepts tend to be vague and imprecise rubrics; to combat the tendency, this chapter will sketch out in some detail what topics of study are brought into focus through the concept.[7]

"Epistemic modernization" refers to the process by which the agendas, concepts, and methods of scientific research are opened up to the scrutiny, influence, and participation of users, patients, non-governmental organizations, social movements, ethnic minority groups, women, and other social groups that represent perspectives on knowledge that may be different from those of economic and political elites and those of mainstream scientists. In a sense the change represents a return, but under very different historical circumstances, to the conditions of early modern science. In the history of early modern science, an epistemic primitive accumulation occurred when Western explorers and scientists traveled around the world and brought home the diverse local knowledges of plants, animals, landscapes, languages, medicines, and social institutions. Modern science as we known it today was built up from the interaction with and codification of lay and non-Western knowledges. As science became increasingly denaturalized and laboratory based, the lay and local knowledges became less important to and less valued in most scientific fields. However, in the late twentieth century various criticisms of science reopened the doors to greater public interaction. Epistemic modernization has echoes of an earlier era but is also quite different.

The epistemic modernization of science and technology occurs as a result of various institutional changes. In the simplest form, the social composition of science is undergoing differentiation both within societies, as the doors are opened up to previously excluded social groups, and internationally, as university-based research and education becomes ubiquitous in countries around the world. Another form of epistemic modernization is the development of community-oriented research projects that involve laypeople in the tasks of agenda setting, problem definition, research design, and implementation. A third form involves the growth of interaction with civil society and social-movement organizations that have increasingly challenged the epistemic authority of science. Finally, a fourth form comes internally, from dissident scientists who have broken rank with consensus opinion over agenda-related issues. The dissidents sometimes have coalesced into NGOs of socially responsible scientists, formed coalitions with NGOs and social-movement organizations to provide them with counter-expertise, and contributed to the development of alternative research fields.[8]

Together the challenges to consensus knowledge and research priorities put into question the policy principle of exceptionalism. An expression of the autonomy assumption as it is translated into science policy, exceptionalism holds that public oversight of scientific research should be an exception to the general pattern of democratic oversight that underlies most publicly funded institutions. The policy principle can be maintained only by convincing the public of the value of professional autonomy. Scientists argue that the output of scientific research and technological innovation is generally beneficial to society (as in the case of advances in medical knowledge); at the same time, they argue that new knowledge was a source of industrial innovation: what was good for science was good for America (or France, or Brazil, . . .). However, as perceptions of risks and hazards generated by science and technology have mounted, and as the public has become more aware of how cultural baggage and industrial interests that are not necessarily identical to a broader public interest have shaped science, the exceptionalist policy has been increasingly difficult to maintain.[9]

The decline of exceptionalism often results in conflict over the framing of science with respect to a broader public interest. Groups that claim to represent the public interest from within science, such as leaders of research fields who serve as spokespersons for good science, adopt the frame of paternalistic progressivism; that is, they wrap themselves in the empiricist flag of gradual scientific progress and methodological neutrality, and they tend to reject as unscientific the knowledge claims of dissident scientists, laypeople, social movements, or reform movements in science that are aligned with social movements. In contrast, those who support alternative research agendas, such as civil-society organizations and reformist scientists, often adopt a corresponding frame or discourse of scientific devolutionism, that is, a historical narrative of a fall from grace in which scientific research has been captured by corporate profit motives and corrupted by cultural bias.[10]

The emergent institutions of epistemic modernization represent a third approach, which attempts to engage the loss of confidence in the equation of "Science" with a general or public interest. Rather than accept the devolution frame or attempt to counter it with the frame of paternalistic progressivism, research under conditions of epistemic modernization accepts the challenges mounted by dissident experts and civil-society organizations but incorporates and transforms them through an emergent institutional structure that opens up scientific inquiry, and to a degree technological design, to participation by non-experts and historically excluded groups. Rather than reject all epistemic challenges as unscientific, the approach selects some challenges for research and some challengers for research funding. In the sections that follow, I will consider four dimensions of epistemic modernization: the increasing diversification of the social composition of science, community-based research, the rise of the interactive model of public communication, and alternative pathways in science.

Diversification of Social Composition

In its simplest and most straightforward form, epistemic modernization results from the effects that the increasing universalism of the membership policies of science as an institution has on its research agendas. Especially since the middle of the twentieth century, there has been growth in the opportunities for women, African-Americans, Latinos, gays, lesbians, and other historically excluded social groups of people to gain access to education and jobs in science. At the same time science has become more internationalized, so that graduate programs and major universities exist around the world and in many postcolonial societies, where there is now the option to pursue graduate work at home rather than in a former colonial center.

One might argue that because scientific knowledge production is (or should be) demographically neutral, such demographic changes should have no effect on the content of scientific knowledge. The argument is defensible up to a point, and at an individual level the science of new members of formerly excluded social categories may appear to be indistinguishable from that of white males in North America and Europe. The argument can also be taken to considerable length when applied to methods and the knowledge vetting process, even across international communities. American scientists sometimes dismiss the research of foreign scientists as methodologically unsophisticated; however, when international controversies heat up, arguments that allude to the demographic address of a scientist have the same illegitimate status in the rhetoric of science that ad hominem arguments have. Demographically based prejudices must be screened and translated into methodological arguments in order to count as legitimate.

However, if one accepts the general argument outlined in the previous chapter, then the theories, concepts, methods, and problem areas of a specific research field are not universally shared but instead are contested between dominant and non-dominant networks. In some cases the non-dominant networks may not be demographically different from the dominant networks, but one would also expect to find cases where the social address of challenger scientists is closely linked to controversies that have emerged. As Donna Haraway has shown for primatologists and Sharon Traweek for physicists, social differences can translate into epistemic differences regarding preferences for methods, problem areas, concepts, and even equipment design. Furthermore, when one examines how the diversification of the social address of scientists has affected the agenda-setting politics of scientific knowledge (that is, the choices of which research fields to develop and which ones to leave undone), the issue of demographic change is hardly trivial. When the doors of science open to a broader social composition, the answer to the question of what counts in science as an important problem area, especially for health and environmental research, depends a great deal on whom one asks. Debates over different agendas and methods benefit science in the sense of making visible unseen biases that have previously passed as unquestioned neutrality. The debates also result in improved research methods and better allocation of resources to problem areas. To use Sandra Harding's phrase, diversity leads to "stronger objectivity."[11]

A parallel process occurs in the technological and design fields (such as engineering, marketing, and the design professions), where technology and product development teams have also opened up innovation processes to diverse social and occupational addresses. By increasing the social diversity of the design team, new opportunities emerge for seeing potential solutions from previously excluded and invisible perspectives. The diversification of the social composition of the professional design teams can be enhanced through methods such as participatory design, which are analogous to the institutions of lay interaction in science to be discussed in the next sections. As a result of both increased professional diversity and the incorporation of lay perspectives in the design process, design teams are better able to check their own views of what users want and how users interact with the prototypes. The innovations that follow are also likely to be more universal. Universal design is to technology as strong objectivity is to science.

In following the research of feminist and multicultural science studies scholars—who argue that changes in the social composition of science lead to the modernization of science in the sense of making possible more robust and universalistic conceptual systems, methods, and prioritization of problem areas—it is important not to overstate the case and thereby to discredit it with essentialism. There is not necessarily an identifiable woman's perspective or African-American perspective on every scientific issue and method. Likewise, the transnational nature of scientific conferences and journals may have to some degree reduced some of the more dramatic cultural differences in science, such as the differences between French and English styles of physics that Pierre Duhem noted at the beginning of the twentieth century. However, the scientific field has become more diverse socially, and there are many instances of networks of scientists who belong to historically excluded groups and bring a new sensitivity to what counts as science. In doing so, they contribute to the modernization of science in the sense of the ongoing critique of cultural baggage that has previously passed as neutral and universal knowledge, methods, and priorities for research.[12]

Community-Oriented Research

In the second dimension of epistemic modernization, the scientist-layperson relationship is brought into close contact through local, community-oriented research projects that involve lay participation. One example is pro bono community-oriented research, in which a researcher undertakes a project without extramural funding and with the intention of benefiting the local community. Self-funding allows the researcher great flexibility in defining the extent of public participation, but in most research fields the absence of funding also restricts the choice of methods and scope of the project.

The pro bono type of community-oriented research was institutionalized in Europe through the development of science shops—small offices that are funded by a university to serve as conduits between the research needs of the surrounding community and the resources of the university. Under the science-shop model, a community-based organization comes to the university office with a research problem, and the office attempts to find a faculty member who is willing to work on the problem or to supervise students who work on the problem. The university generally provides some limited funding to staff the office, such as by paying for a part-time staff person, but the research is conducted by faculty members or students on a pro bono basis. The "science-shop movement" (as it was sometimes called) grew during the 1970s and the 1980s, but it suffered retrenchment and cutbacks under the budgetary pressures and the neoliberal reorientation of the universities in the 1990s and after.[13]

Universities in the United States never developed a similar movement of science shops in the sense of a university-funded office that recruited and sorted through proposals from community groups in order to link them to pro bono research projects completed by faculty members and students. Instead, the idea of science shops has tended to develop piecemeal as individual researchers bring scientific expertise to community groups or as service-based learning programs connect students with community organizations. Some universities also have centers oriented toward community development, but the centers offer a more restricted range of research expertise than the European science shops.[14]

Somewhat analogous to science shops, and more common in the United States, is community-based research. In the science-shop model, the community organization is involved in preliminary articulation of the problem, but it is not necessarily involved in the design and execution of the research. For controversial issues, scientists may even prefer to insulate themselves from additional participation from the community group so that the results of their research do not seem tainted by participation from one side of a controversy. In community-based research, problem definition, design, and execution of the research are more often the outcome of a collaborative process. Community-based research draws on the deeper theoretical and practical research traditions that are usually described as action research and participatory action research. The research tradition has been traced back to the collaborations between sociologists at the University of Chicago and Hull House during the 1890s and to Kurt Lewin's work on race relations during the 1940s. After the 1960s, Paulo Freire and other researchers associated with the liberation struggles in developing countries developed a more activist wave of action research.[15]

Community-based research may originate from the community or the university. An example of the former is the emergence of what Phil Brown calls "popular epidemiology," in which residents in a community identify a new disease or a cluster of cases of a known disease, and they may also identify potential causes. In the case of the cancer cluster that developed in Woburn, Massachusetts, the community members also sought out and developed a partnership with researchers at Harvard University's School of Public Health. In a somewhat different example, researchers at the University of Pennsylvania developed partnerships with various community groups in the impoverished surrounding community of West Philadelphia. The university provides support for faculty members and students to engage in various participatory-action projects, many of which take place through the local schools, such as a nutrition education project.[16]

To the extent that the university defines its core mission as an engine of technological innovation oriented toward regional industries, it will be increasingly difficult to find a place for pro bono research projects that involve the participation of community organizations and/or social-movement organizations. There will be exceptions, such as when universities are located in low-income neighborhoods and find it in their enlightened self-interest to support community-development projects and partnerships with community organizations. Even in those cases, the models of research associated with business-oriented economic development (the "right hand" of the university) are likely to receive much more attention and greater resources than those driven by participatory action. The growth of service learning programs may represent one opportunity for pro bono community-based research, but the quality of research that is possible in short-term internships is unlikely to serve the community's needs for collaborative research and expertise.

One opportunity for growth for community-based research, and in many ways the model for this type of epistemic modernization in the United States, is public health research oriented toward low-income, ethnic minority, and/or rural populations. The history of clashes between researchers and community groups regarding research agendas, methods, and bias has motivated health and environmental funders to be more concerned with disadvantaged populations and much more sensitive to community access and participation. For example, in 2001 the Agency for Healthcare Research and Quality, in collaboration with the Kellogg Foundation, held a conference that developed recommendations on community-based participatory research. Likewise, in 2005 the U.S. Environmental Protection Agency supported a similar conference on community-based participatory research and environmental justice. Those changes represent recognition by mainstream governmental and funding organizations of the value of enhancing community access to research agendas and research.[17]

The Interactive Model of Public Communication

Another dimension of epistemic modernization has been the shift in the public communication models of scientists, engineers, medical researchers, and other expert groups. The transmission model, which was associated with the exceptionalist policy for science and the assumption of the value of scientific autonomy for a democracy, defined communication between scientists and the public as a one-way process that took place through the media and the educational system. In turn, feedback generally came not through the media and the educational system but through passive public support of taxpayer assistance for research dollars authorized by elected public officials or through contributions to non-profit research organizations.

Associated with the paternalism of the transmission model is an internal policy of suppression of the dissident expert. Suppression can include employment campaigns to subject challengers to dismissal, funding cuts to challenging research networks, and media campaigns and litigation to discredit and exhaust challengers. The worst suppression may be reserved for the high-status insiders turned critics and challengers. As the news reverberates among scientists who may be sympathetic to the dissidents, suppression creates a chilling effect for other would-be sympathizers and challengers. The suppression strategy tends to be correlated with the ideological response of paternalistic progressivism, in which there is a blanket rejection of challenging knowledge claims, and second-tier philosophers may be encouraged to legitimate the labeling of pseudo-science. However, the evidence of bias and suppression feeds the radical critiques of NGOs and civil-society organizations that adopt the scientific devolution frame.[18]

In addition to discrediting internal dissidents, the leaders of scientific research fields also discredited lay knowledge in general by developing a survey literature on the public understanding of science. The surveys document a lack of general public knowledge about basic science, and they shore up the claim in favor of funding and policies that support enhanced transmission of scientific knowledge to the general public. However, the survey literature is generally silent on other forms of ignorance, including, as Brian Wynne has noted, scientists' misunderstanding of the public and lay knowledge. Alternative surveys that might reveal the "shocking" lack of literacy among scientists, engineers, doctors, and other technical professionals about lay knowledge (for example, the experiences of workers with workplace hazards, community perceptions of environmental and health risks, patients' understanding of their illnesses, or other areas of lay knowledge) remain undone science. The documentation of the public's scientific illiteracy can be helpful to scientists who wish to increase support for transmission-related programs that utilize the media and the educational system, but it tends to support a one-way model of communication.[19]

In contrast, a growing body of literature on the public understanding of science, which interestingly relies on the alternative social-science methodology of qualitative interviews and ethnography rather than quantitative survey methods, has shown the remarkable ability of lay publics to acquire expertise and fluency when necessary, as in the case of communities that face environmental justice issues or patients who battle chronic disease. The literature documents the ability of publics to reconstruct knowledges and reappropriate technologies for their own purposes. Furthermore, it tends to articulate a concept of the public that is more differentiated, with pockets of literacy and illiteracy that are strategically based on a need to know. The pockets may take the form of geography-based local knowledge, experience-based knowledge (such as that of chronic disease patients), or combinations of the two (such as communities that are suffering the health effects of toxic exposure). Although the transmission model dismisses their knowledge as scientifically groundless, they operate from an alternative epistemic authority grounded in personal experience. Their own variants of "Cogito ergo sum"—"I am sick; therefore, I doubt," "I can smell the pollution in the air; therefore, I doubt," and so on—provide the basis for their confidence in questioning expert authority and their need to engage it in detail.[20]

Under conditions of epistemic modernization, the assumptions behind the transmission model are put into question, and an interactive model emerges in its place. Research communities come to recognize civil-society groups, such as environmentalists and disease-based patient advocacy groups, as bearers of legitimate questions rather than merely a lack of knowledge and misinformation, and they recognize lay knowledge as complementary to scientific knowledge rather than merely inferior to it. Scientists who embrace epistemic modernization replace the older communication policy of improved transmission to the lay masses and stepped-up suppression for the insider challengers with the institutionalized incorporation of selected epistemic challenges. The challenges can be converted into researchable knowledge claims, just as some challengers can be converted into institutionalized participants in research and in policy making.

The changes are especially evident in the incorporation of health and environmental social movements. What once took a social movement to open up research policy making to public debate (AIDS, breast cancer, alternative cancer therapies, anti-toxics, etc.) is now increasingly institutionalized through the conversion of social-movement organizations into insider advocacy organizations with their own affiliated research fields. As research undergoes epistemic modernization, the public's involvement in and shaping of science becomes not only more prevalent, but recognized by the leaders of the scientific community as a legitimate part of the agenda-setting process. As Brown has noted, citizen-scientist alliances also form to challenge the accepted scientific knowledge. To the extent that the change becomes widespread and exercises a significant influence on research agendas, we can say that the knowledge-making process has undergone epistemic modernization. Again, the question can be studied empirically across scientific fields.[21]

There are two major forms in which the interactive model is emerging. In the indirect form, interaction involves contributions by lay advocates and activists to processes that set overall agendas. Usually the work involves an offshoot of social-movement or activist groups in which some activists have developed the appropriate literacy to engage the legislative appropriation and public funding agencies; in other words, they have undergone the expertification process described by Steven Epstein. The highly qualified lay knowledge is not equivalent to the expertise of scientists; it involves what I call "narrow-band competence." In other words, the activists are competent in a narrow range of the relevant science, but they tend to lack the broader knowledge of a technical discipline or a research problem area that is typical of scientific researchers. Consequently, they can fall off the knowledge cliff of competence in an expert field of knowledge fairly easily, and their contribution tends to be restricted to the broader issues of funding allocation and agenda setting.[22]

Notwithstanding the shortcomings, activists and advocates can become competent enough to engage the networks that shape funding, and, if successful, they can contribute to shifts in funding priorities and research agendas across or within research fields. In some cases, interaction with activist and advocacy organizations has been institutionalized through positions on government agency funding panels. An example can be found in the advisory boards of the National Institutes of Health, where patient advocacy groups have won seats at the table of funding decisions. The outcomes of participation may in some cases reflect advocacy goals for shifts in funding agendas; however, research communities can also ignore lay viewpoints. Likewise, when activists have achieved funding shifts through direct lobbying of legislators, the research communities can recapture and rechannel the funds.

Another institutionalized example of lay participation in agenda setting is the consensus conference. The Danish institution utilizes the model of the jury trial, with random selection as the democratic principle, to develop a citizen panel of non-experts. With the support of expert teachers, the panel deliberates and provides advice on a policy issue such as whether a research field should be pursued and, if so, what regulatory guidelines should be put in place. Although consensus conferences can be set up to include representation of civil-society groups, they have tended to be set up around the individualist assumption of citizen participation. In the United States there have been only a few experiments in the use of the Danish consensus conference model, but they have not been linked to the policy-making process, and their effect on the goal of increasing lay access to scientific and technological decision making has been limited. More likely to effect policy changes are the summits and informal meetings that sometimes occur among regulators, NGOs, and research leaders.[23]

In the direct form of the interactive model, activists and advocates make the transition from contributors to general agenda setting and formulation of regulatory issues to the status of contributors to scientific research fields. In my historical and ethnographic research, three basic processes became evident: conversion, biographical transformation, and network assemblage.[24]

In the first type, activists target scientists who are working in a dominant research program and attempt to convince them to shift to a research program aligned with the activists' goals, and they provide the scientists with access to information and collegial networks. In the environmental movement, conversions appear to be more easily accomplished with emeritus and retired researchers, who are no longer dependent on the funding pipeline and who can therefore afford to take the risks of operating outside the mainstream. The scientists need not engage directly in new research; they may also provide counter-expertise, such as advice to the activist organizations or expert testimony in their support.[25]

The second type, biographical transformation, represents an extreme variation of the expertification process: the change of activists or advocates into researchers through additional formal education and credentialing as scientists. The change of status may involve acquiring the knowledge to be able to participate as contributors to research programs, or what Harry Collins calls "contributory" expertise. In that case the direct form of participation can overlap with the first dimension of epistemic modernization, where the social composition of science changes as new social categories gain the credentials to participate in a scientific field.[26]

In the third type, network assemblage, the lay activist or advocate does not necessarily acquire the formal credentials or expertise to contribute to a specific scientific field. However, the advocate can be a catalyst for research projects by obtaining funding, by setting up a foundation, or even by leading the research projects but delegating the more technical work to experts. One form of contribution is the review essay with a byline of mixed contributions from advocates and scientists, which can be published in peer-reviewed journals but does not require new research.[27]

In summary, the demise of the transmission model entails recognition from the leaders of science that a new kind of relationship with the public is possible, one that is analogous to the transition in the media from broadcast to interactive communication. Unlike community-based research, in which the interaction with laypeople is more localized through community organizations on specific research problems, the interactive model is less geographically localized and more oriented toward general research agendas. The lay activists or advocates can find places on funding panels and conferences, help scientists to convert to new research programs, undertake further education to become experts themselves, arrange for funding and research opportunities for sympathetic scientists, and help orchestrate research projects.

Alternative Pathways in Science: Opposition

The fourth major dimension of epistemic modernization involves the way in which scientists themselves directly incorporate into their research programs the social-change goals associated with social movements. Usually, the dissident scientists utilize the repertoires of action available within the scientific field; consequently, this type of mobilization is best described in the terminology that I have developed as a professional reform movement within science. However, the scientific reform movements often exist alongside and in alliance with broader social movements, so there is considerable overlap between reform movements within the scientific field and social-movement mobilizations. This section and the next will discuss some of the alternative pathways in science that emerge in two types of reform movements within science: those with a goal of stopping a field of research and those with a goal of developing an alternative research field. The discussion here will be parallel with the similar discussion in chapters 4 and 5 for alternative pathways in industrial fields.[28]

One way of engaging the politics of research agendas is to call for a moratorium on certain areas of research. In the United States, since the middle of the twentieth century, the most consistent area of scientist-driven mobilization to stop an area of research has been war-related research. Kelly Moore's work charts the radicalization of scientists during the late 1960s around the anti-Vietnam War movement and also traces their roots to a longer history of scientists' peace activism that dates back at least to the middle of the twentieth century. As the anti-war movement heated up, various radical and public interest science organizations opposed the development of specific weapons and urged colleagues to sign pledges not to undertake weapons research. At Stanford University and at the Massachusetts Institute of Technology, protests against weapons-related research resulted in the severing of ties between universities and weapons laboratories. Scientists also opposed President Richard Nixon's plan for the development of an "anti-ballistic missile." A second wave of anti-weapons protests occurred during the 1980s, when scientists opposed the Reagan administration's plans for a "Strategic Defense Initiative." In both the earlier and the later protests, scientists signed pledges not to engage in weapons research.[29]

Another area in which some scientists have advocated a research moratorium is biotechnology research. A controversy erupted in 1971 over plans to implant a bacterium with DNA from the tumor virus SV40, but concern within the scientific community soon became general, and a movement arose among scientists to regulate the emerging field of biotechnology research for safety purposes. As the public became more aware of the controversy, calls for government regulation increased, and in 1977 the city of Cambridge, Massachusetts initiated a wave of local and state government moratoria on high-risk types of recombinant DNA research. Several bills were introduced in Congress to limit DNA research, but they were not passed, and instead guidelines developed by the National Institutes of Health resolved the controversy.[30]

Unlike the weapons-related pledges, in the case of recombinant DNA research the intention of the scientists was not to halt the development of a research field but only to put in place safety guidelines and to put some restrictions on the riskiest research. When the research moratoria did emerge, they came from outside the scientific community in the form of temporary policies imposed by local and state governments. The pattern recurs in subsequent research moratoria in biotechnology directed at research on cloning and stem cells, where the government has imposed moratoria, albeit due more to moral considerations than an assessment of risk. In all three cases (recombinant DNA, cloning, and stem cells), concerns and pressure external to the research community drove the process, and the government-based moratoria have tended to be ineffective. In the case of recombinant DNA, once the research community developed a consensus around safety standards the research and eventually the associated industry could proceed. Likewise, even in the absence of federal funding for stem cell research, the field continued to grow, with state and private funding. More generally, one might hypothesize that if the federal government were to cut off research funding for a field, but industry were to remain interested in developing the field (as in the biotechnology cases), the research would proceed even in the absence of public funding. In the event of a complete national moratorium not merely on funding but on the research itself, scientists could go to other countries, and from there they could weaken the moratorium by pointing to its conflict with national competitiveness goals.

The situation is somewhat different in the case of weapons research, where the drive for a moratorium came more from the scientists themselves, and it took the form of a pledge not to engage in a type of research. In effect, scientists had called a research strike. The pledge represents only the organized end of a continuum of responses to research programs that are controversial to segments of a research community. At the other end of the continuum is the stigmatization of research programs that benefit the military or a specific industry on the grounds that the research would generate technologies that create undue risks for the broader society. For example, scientists who work on genetically modified food for agribusiness, on nuclear or fossil-fuel energy, on chlorine-based chemistry, and on related topics that are environmentally controversial find that their research programs can be the target of criticism from environmental groups as well as stigmatization from some of their colleagues. An example of organized resistance is what Frederick Buttel has termed "Hightowerism," referring to the failed mobilization during the 1970s to attempt to stop public agricultural colleges from engaging in agribusiness research.[31]

The cases of an organized pledge and stigmatized research fields are examples of how broader public concerns, as articulated by social-movement or advocacy organizations, can be translated into the scientific field in the form of debates and mobilizations by scientists in opposition to some research agendas. They provide another example of epistemic modernization, in that there is contestation over the contours of a research field as determined by political and economic elites. To date, however, mobilizations by scientists to stop the development of a research field have not been particularly successful. As long as funding continues to flow for a research field, there are likely to be scientists available who will accept the funding. It is difficult to maintain solidarity in the ranks of a loosely organized, entrepreneurial profession when huge pools of military-industrial funding become available. Some scientists will select the opportunities even at the risk of stigmatization from colleagues. Because decisions about research programs are individualized, individual scientists may opt not to accept certain categories of funding (such as from the military or environmentally controversial corporations), but others are likely to fill the gap. As a result, if an internally generated moratorium is to be successful, scientists must go beyond the pledge by lobbying Congress to end research funding.

If a moratorium emerges from the scientists in the form of a grassroots pledge (such as to refuse to undertake weapons research) but the scientists do not persuade the government to end funding, the pledge will be weakened by "scab" scientists who opt to accept the stigmatized funds.

Even if the scab scientists come from less prestigious universities, and the high-prestige scientists are able to maintain solidarity among their ranks, the new flows of funding, if significant enough, can begin to shift the prestige ranking of the universities. One can see the general effect of funding on prestige by looking at some American universities that were relatively unknown at the beginning of the twentieth century and have grown in prestige by developing significant military and industrial research patronage. If concerns with shifts or erosion of prestige emerge, or even if departments face the loss of significant sponsored research funds, administrators may "correct" a research strike by imposing external reviews on the department, by squeezing resources, or even by rechanneling resources into new, independent research centers that are aligned with strategic goals and circumvent departmental control.

Alternative Pathways in Science: Non-Dominant Research Fields

Oppositional politics within the scientific field can coincide with a second type of scientific mobilization: the development of alternative research fields. For example, the "radical science" movement of the 1970s diversified from opposition to war-related research and technology to various projects in support of "people's science," which included agricultural, computing, and scientific assistance for North Vietnam and Nicaragua as well as for underserved segments of the population in the United States. The experiments in people's science shifted activism in science away from the oppositional politics of the research moratorium toward new research programs that were geared to the needs of the world's poor, and toward the idea of developing alternative knowledges and technologies. In turn, the goal of developing sciences for the people drew on and supported the appropriate technology movement, a grassroots research and development movement that emerged in non-profit organizations largely outside the major universities. The appropriate technology movement drew attention to the goal of developing new energy and agricultural technologies that today would be recognized under the rubric of sustainable design (particularly the localist strands discussed in chapter 6). Although there are many types of alternative research fields, I will focus here on three that are relevant to environmental issues: renewable energy, sustainable agriculture, and green chemistry.

I will briefly flag some of the issues regarding alternative pathways in science, then return to them in more detail in chapter 5.[32]

In all cases, the alternative scientific fields oriented toward more environmentally sustainable science show significant but limited gains since the 1960s. Support from the federal government has opened up for each of the three fields. For example, research on renewable energy achieved recognition during the 1970s in the wake of the rise in oil prices and the decline of public confidence in nuclear energy. At that time, the federal government earmarked some funding for research on "soft path" alternatives such as solar energy. In the 1980s, the federal government launched what later was named the Sustainable Agriculture Research and Education program. In the 1990s, the Environmental Protection Agency began supporting research initiatives in green chemistry.[33]

In each case the fields can point to some success stories. Renewable energy has become increasingly mainstreamed through industrial growth in areas such as wind and solar energy. As the industry has developed and government-based funding has become available, there has been corresponding growth of scientific journals and research networks as well as support for research and development in the private sector. Even in times of retrenchment of federal research spending, such as during the presidency of George W. Bush, renewable energy research has received some funding and policy support. Likewise, sustainable agriculture research can point to some growth of acres under organic cultivation at research universities and to the development of sustainable agriculture research centers in some of the major agricultural schools. Green chemistry can also point to the growth of research panels at mainstream chemistry conferences and to the emergence of conferences dedicated to "green chemistry."[34]

However, when set against the broader background of funding levels in comparable fields, the growth of the three research fields must be understood as having occurred in a context of low levels of relative funding. For example, federal funding for renewable energy research and development peaked at about $1.6 billion in 1979, declined throughout the 1980s, and in real dollars never returned to the level of support of the late 1970s. Across both Democratic and Republican administrations since the 1970s, nuclear energy research and fossil fuel research have each consistently received more funding than renewable energy.

Furthermore, funding for renewable energy has been highly selective; in the early 2000s, the presidential administration favored "road maps" for a long-term transition to the hydrogen economy and ethanol fuels but fired (at least temporarily) a number of researchers at the National Renewable Energy Laboratory.[35]

A similar pattern of relative lack of research funding can also be found for organic agriculture and green chemistry. A search on the U.S. Department of Agriculture's database of research projects in 1997 found that only 34 projects out of 30,000 were explicitly directed toward organic farming systems and methods (about 0.1 percent of the database and of Department of Agriculture funding). Likewise, a study in 2003 found that organic agriculture research acreage was only about 0.13 percent of total research acreage. If anything, the studies overestimated the total proportion of research on organic farming because they focused on government sponsored research; industrial research funding has overwhelmingly favored biotechnology and other types of conventional agriculture. Another sign of the weakness of the research field is the fact that researchers interested in career advancement tended not to select sustainable agriculture as a research field. Just as most federal spending for energy research is devoted to fossil fuels and nuclear energy, most agriculture research is dedicated to biotechnology and to increasing monoculture production through industrial agriculture. A similar pattern of marginal funding plagues green chemistry. No major university's chemistry department or chemical engineering department is dedicated to it, and research programs in green chemistry tend to be viewed with disdain because they are seen as "too applied." Federal funding for research in this area has been on the order of 1–2 percent of the total funding of $300 million for chemistry research.[36]

From the optimistic perspective, research on renewable energy, organic agriculture, and green chemistry are examples of growing alternative research fields that attempt to fill in the gaps of undone science. However, to the extent that scientists who sympathize with environmental movements wish to undertake research in the three areas, they will be joining non-dominant research networks and facing the attendant career risks. The risks are probably greater for agricultural scientists who opt to work on organic research and chemists who opt to study green chemistry than for engineers who want to work on solar power or other forms of patentable green technology. Since the late 1990s renewable energy has become a growth industry, and in the early 2000s there was a wave of solar energy startups that recalled the 1990s' wave of biotechnology startups. As a result the research field has undergone a transition from the alternative status to alignment with competitiveness and industrial development goals. Individual researchers who make a decision to develop a patentable product and become an entrepreneur have the option of becoming wealthy, but in essence they are also leaving the prestige game of reputation building through publication.

Whatever the shortcomings of the alternative research fields, the fact that they exist and have shown some growth is, like the existence of attempts by some scientists to end some categories of weapons research, another indication of the permeability of the scientific field to shifts in agendas that are consistent with the goals of the activists and advocates such as environmentalists. The alternative pathways in science, and their linkages to alternative pathways in industry, represent another dimension along which the epistemic modernization of science can be tracked.

Conclusions

Although conflict and controversy among networks in a research field are inherent in the scientific field, two historical changes associated with globalization enhance the level of conflict. First, there is an increasing emphasis on mission-based funding oriented toward technology transfer and industrial innovation, especially when geared to national industrial priorities and regional industrial clusters. As a result the selection of research agendas becomes a policy problem to be addressed from the value perspective of the competitiveness of industrial (and military) innovation. Second, there is a countervailing trend of epistemic modernization, which involves opening up the content of scientific research fields to greater public participation and influence. In some cases conflicts within a scientific field between two research networks are parallel to general societal conflicts, where two articulations of a public benefit (one defined by military-industrial organizations and one defined by civil-society organizations) are in conflict. However, because there is always some degree of autonomy in the scientific field, the relations of cooperation and conflict among networks in science do not always map neatly onto broader social divisions.

Even where the alignments with broader social divisions are not readily visible, funding priorities shape network dominance. Because the priorities of funding sources reflect, however imperfectly, the interaction of the negotiated priorities of scientists and those of economic and political elites, there tends to be an alignment between the dominant networks of a research field and the interests of the elites. Researchers who are developing transferable and licensable technology will tend to win huge helpings of funding served on elegant platters, whereas those who wish to explore the health and environmental effects of such technologies may end up being sent to the kitchen to beg for the scraps off the table of the funding system. Because dominant networks tend to control access to the means of disciplinary reproduction (journals, departments, graduate students, and funding panels), they can afford to ignore the non-dominant networks and let them wither on the vine of inattention. No conspiracy theory is needed to explain the alignments that occur; one needs only to understand that the fields of science are not autonomous regarding the self-determination of the broad priorities of research agendas. To some degree they never were: from the seventeenth through the twentieth century scientists have always fought to maintain a degree of autonomy from extra-scientific intervention. Attention to the specific institutional changes that have occurred in the era of globalization makes it possible to understand how the scientific field is increasingly a site where general societal conflicts play themselves out.

In view of the ongoing influence that industrial and political elites exercise over research agendas, not to mention the industrial processes for technological innovation, epistemic modernization could be viewed as little more than a strategy to colonize lay knowledge, co-opt civil-society challengers, and quell internal dissidents and reform efforts. In discussing the phenomenon of epistemic modernization, one should not forget that the dominant changes in scientific fields will be driven by the dominant groups in society. Although the quasi-autonomy of the scientific field also makes some exceptions possible, this fundamental "law" in the sociology of knowledge has been recognized since its discovery by a marginalized social scientist more than 150 years ago, and it is unlikely to be repealed any time soon. However, the concept of epistemic modernization draws attention to another dimension of a non-autonomist approach to the scientific field in an era of globalization: the role of pressure and participation from below. As with reflexive modernization and ecological modernization, the question of the political significance or lack of significance of epistemic modernization should be left open to empirical research. Research on the topic is likely to result in nuanced determinations about occasions where the processes of epistemic modernization, including alternative pathways in science, are more and less significant.[37]

3

The Transformation of Technological Fields

The emerging alternative technologies and products that address environmental problems, such as renewable energy and organic agriculture, flourish and languish in broader technological fields. Like scientific fields, technological fields consist of relations of competition and cooperation among networks, in this case centered on material objects rather than research programs. Also similar to scientific knowledge, new technologies and products are subject to a cultural politics and political economy of selection that determines which ones flourish and which ones are left in the demonstration-project stage. Yet there are also significant differences. For example, the claim that technologies, products, and other aspects of material culture are socially shaped is perhaps banal and certainly less controversial than a similar claim for scientific knowledge and methods. From this perspective the theoretical task is less difficult. However, there are still assumptions about the dynamics of technological fields that should be left behind.

Whereas the analysis of scientific fields began with the widely held view of scientific change based on a simplistic empiricism, the analysis of technological fields begins with the widely held theory of technological change based on a simplistic view of efficiency. This chapter will develop a critique of the theory of technical efficiency as the only significant driver of technical change, then it will develop a broader framework for the analysis of technological fields that is parallel to the analysis of scientific fields developed in chapters 1 and 2. Before developing the framework, I will set forth some basic definitions.

In the broadest sense technology can be understood as a product, infrastructure, tool, machine, or other form of material culture that has been designed in order to assist human activity and to make changes in the material and/or social worlds. We generally think of technology in more narrow terms as a machine that has been designed to assist human activity. The narrow definition has its place, but the boundaries between machine and tool, or technology and other forms of intentionally designed material culture, are by no means clear. For the purpose of developing a general model of technological change, it is less important to draw sharp distinctions between technologies such as agricultural harvesters and tools such as hoes, or likewise between technologies such as buses and infrastructure such as highways, than it is to develop a general framework for the analysis of a field of material culture, which has a history and politics characterized by contours of dominant and non-dominant technologies.

To understand the differences among technologies within a technological field, categories such as machine versus tool are less valuable than the concept of design differences, which can occur within a given category of technology, such as among types of transportation technologies or agricultural technologies. The design of a technology, product, or other unit of material culture is the result of the intentional application of techniques, sometimes with assistance of tools and machines, to shape and differentiate it. Differences in design can be mapped not only to differences in the efficiency of a technology or product, but also to differences among social categories of producers, users, beneficiaries, and victims. As Marshall Sahlins noted, a totemism of material culture, such as clothing styles that make statements about social identities, is fundamental to modern and pre-modern societies alike. Design differences can mean a lot to industries, firms, social classes, genders, user groups, and ethnic groups. The alignment of material differences with social differences is the basis for the politics of technology, because the selection of some designs over others usually affects both social categories and the biophysical environment differentially. Even if one accepts that there is always interpretive flexibility in material culture and that users can reconstruct and reappropriate it, Langdon Winner's phrase "Do artifacts have politics?" still points to the ways in which design choices can restrict potential uses and users. By beginning with the understanding that design decisions and technological choices have a social and political dimension, the stage is set for developing a critique of technical efficiency as the primary driver of technological change.[1]

The Limitations of Efficiency Explanations

If one goes to a museum and looks at a display of fire engines or airplanes that have been arranged to show a temporal progression, or if one attends the "Carousel of Progress" at Disney World, technological development is ordered historically to show how machines have improved in their role of assisting humans with work. If one analyzes what improvement means, it can usually be equated with increased efficiency in the sense of speed or capacity. Improved speed and capacity will generally also save energy and labor, as in household appliances and industrial machines that reduce the human effort and time needed to accomplish a task. The relationship between new and old technology is characterized as one of subsumption. In other words, the new technology must meet the same goals but with greater speed and/or enhanced capacity, or it may have the same speed and capacity, but it should reduce energy and/or labor expenditures.[2]

Under the simple efficiency model, technology is viewed as an autonomous field that develops according to its own internal logic of invention and efficiency. Only cost, a constraint on affordability, intervenes from the outside to place limits on and provide incentives for the diffusion of a technology. However, technical and economic efficiency can be brought together in a calculus of benefit and utility. From the perspective of the efficiency model of technological change, when individuals or societies confront a choice between two technologies, the more efficient technology will win. The train replaces the canal barge, the automobile the horse and buggy, the vacuum sweeper the broom, and the emailed message the posted letter. In most situations the new technology has greater capacity than the old, and it may save labor, time, and energy that make it more profitable and justify any additional costs over the old technology. Under the simple or straightforward efficiency model, individuals and communities may still have a choice between the old and new technology, but those who choose the old, such as the Amish, are dismissed as irrational and condemned to become isolated preserves of backwardness. Inefficiency only enters into this model in the form of path dependency, such as when the sunk costs or dependence on another, existing technology prevent a shift to a more efficient alternative.[3]

Implicit in the efficiency model of technological change is an almost automatic process of technology adoption. Excepting in cases of path dependence, it is assumed that the more efficient technology or product will eventually triumph. If the new technology costs more, it will have to offer suitable gains in efficiency, but if it does so, then eventually it is assumed to win out. The process may take time, and there may be longer and shorter curves of adoption, but there is an inevitable quality to the story of diffusion and adoption of the innovation. After all, an advocate of the simple efficiency model of technological change might ask, who would really want to cross the ocean in an early-twentieth-century airplane let alone a seventeenth-century ship? If the new technology is more efficient than the old, society will embrace the improvement, or it will become backward and risk being overrun by other societies that have developed more efficient technologies, especially military ones.

Much as science studies developed the conventionalist critique of simplistic empiricism, so technology studies has developed an alternative view to the simplistic model of technical efficiency as the primary driver of technological change. Since the 1960s, a large body of research in STS has undermined the assumption of the autonomy of technological fields. We know, for example, that design choices between a gas and electric refrigerator did not take place on purely efficiency grounds but instead were manipulated based on the competing interests of the gas and electric industries; that technologies in development undergo phases of openness to multiple design possibilities advocated by relevant social groups and closure due to social negotiation; and that technologies are embedded in sociotechnical systems that include people, organizations, and material culture. Efficiency criteria still matter, just as in science studies evidence matters and the new theoretical framework must cover a wide range of accepted knowledge in order to displace an old theoretical framework. Likewise, in technology studies the new boat must float in a wide range of accepted user conditions before replacing the old boat. However, there are many possible designs of a new boat that floats, many possible ways of defining improved efficiency, and some fuzziness over the boundaries of accepted user conditions. Is the primary criterion to be the speed of the boat, its carrying capacity, the materials used, its health effects, its safety, or its energy consumption? Moreover, who gets to determine which criteria of efficiency are the more and less important ones? In coming to a definition of what an "improvement" in efficiency means, all sorts of conflicts can emerge, and the politics of criteria permeate an adoption process that might, on the surface, appear to be as autonomous and straightforward as the crucial experiment. Something like the stalemate of empirical equivalence with competing scientific theories can also occur among competing technology designs.[4]

The problem of the flexibility of efficiency definitions and criteria, or more generally definitions of what constitutes a design improvement, is solved by an alignment of efficiency criteria with other criteria, such as who will benefit from the technological change and who will lose. Efficiency criteria such as saving labor can shift power relations among workers along lines of skill, class, and gender. Likewise, shifts in the definition of what counts as a more efficient technology can alter power relations between workers as a whole and managers or between one industry and another. Not only can a new technology change power relations, but power relations can change definitions of efficiency and evaluations of technological design. As Winner noted years ago, the view that science and technology are autonomous, irreversible forces weakens the imagination and the political will to think of technology as legislation by other means. In contrast to the view that sciences and technologies influence society as the genies escape from the bottle of brilliant inventors and university laboratories, one needs an alternative framework that focuses on the politics of the selection of design features and the potential for the democratic control of science and technology. Analogous to the framework developed in chapter 1 for science, we need a model that focuses attention on the politics and political economy of technology selection.[5]

The Dynamics of Technological Fields

The concept of a sociotechnical system goes a long way toward addressing the need for an alternative framework of technological change. As Thomas Hughes noted, technologies are embedded in complex systems that include organizations, inventors, managers, financiers, engineers, users, and regulators. Hughes's work has demonstrated that even for technological systems that have high levels of capitalization, strong momentum, and a friendly regulatory environment, there are frequent challenges for the system's growth and diffusion for which the concept of efficiency is hardly adequate. Instead, he employed the military metaphor of reverse salients, which describe aspects of a technological system that fall behind as the system grows and advances, in analogy with a front line of troops that has been held back by particularly tough resistance. To address the reverse salients and preserve the momentum of a sociotechnical system's growth, system managers deploy technical and organizational efforts, including invention. As he suggested, reverse salients are a source of many, perhaps even most, inventions, even if inventions in this context are not necessarily the most radical.[6]

The challenges are even more formidable when an alternative technological system is competing for position with a long-standing and established technological system, such as occurs for distributed solar energy versus grid-supplied fossil-fuel energy or for organic agriculture versus conventional agriculture. When the challenging technology lacks a strong economic base of powerful champions, as is often the case with design innovations that would enhance the general benefit to the environment or the lower income segments of society, the alternatives sometimes take a long time to acquire the capital needed to resolve their reverse salients. The alternatives can also take a long time to achieve the market share that allows them to drive costs down to a competitive level. To understand the dynamic, where various technological systems are competing and the competition is far from even, one needs an approach to technology that is similar to the one developed in chapters 1 and 2 for science: a historical perspective on the dynamics of technological fields.

The concept of a technological field is intended to draw attention to a broader dynamic than a conflict between closely related technology designs that are advocated by different firms, such as the conflict between alternating and direct current electricity that Hughes analyzed. Instead, the growing sociotechnical system or heterogeneous network of organizations, regulations, users, technologies, and associated products is situated in a larger field in which the changing relations of cooperation and conflict with other technological systems take place. Regarding the growth of electrical grids that supply energy to homes and businesses, one might look at patterns in the growth and decline in a wide range of lighting and energy systems for buildings over 150 years or more. The range of the technological field would include human-powered appliances, passive solar designs for lighting and heating, off-grid and distributed energy systems, and choices in the energy source of appliances such as dryers. A technological field can be analyzed at an aggregate level (e.g., the energy industry in a country) or narrowly with a limited geographical scope (e.g., a neighborhood).[7]

As with the analysis of a scientific field, it is not enough to develop an anatomy of the relations among networks at a single point in time. Instead, the analytical framework should open up the history of the field to make visible the patterns in shifts from one system or network to another. The analysis of a technological field over time reveals a wide range of technical differences, some of which are selected and become embedded in the design of the objects as they are developed and diffused, and others of which are excluded. When one adopts a historical perspective, the design at any given point in an artifact's history is an outcome of the selection from a range of possibilities that reflects contingent historical events. The contingent events result in design differences among the objects of a technological field, and the process of selection may be understood and legitimated with reference to efficiency criteria as well as values and power. Often the latter criteria play a role in highly conflict-ridden decisions that may be later justified with reference to efficiency criteria. Conversely, the process of technological development sometimes also involves choices that few parties are even aware have been made. Even the efficiency factors in technological design are inflected in ways that are aligned with social and political differences, so that openly expressed rationales of neutrality based on economic and technical efficiency may legitimate and obscure deeper political alignments.

By examining the objects of a technological field comparatively (across time and space or across technological systems within a field), it becomes possible to see not only the cultural assumptions inherent in stylistic differences but also the alignments of technical, economic, political, social, semiotic, normative, and other differences in the development of a technology. The analysis of technologies comparatively within a technological field reveals a wide range of design possibilities, many of which did not make it into the current range of designs for an existing technology or product. The analysis can be left at a historical and descriptive stage, which provides an anatomy of a field of alignments of difference over time, but it also can be used to provide a clearer perception of alternatives that are available so that an evaluation of options for future action is less muddled. In other words, the analysis of a technological field can help reveal how some alternative pathways have been lost and how some have been incorporated and transformed into the dominant technological systems of the field. As Winner once asked, "Were there any real alternatives? Why weren't those alternatives selected at the time? How could any such alternative be reclaimed now?"[8]

For example, in the eighteenth and nineteenth centuries the transportation field of upstate New York was critical in American history because it provided a riverine link from the Atlantic Ocean to the great lakes and the interior of the continent. Waves of transportation systems (canals, railroads, roads, and the state's thruway) followed the original network of trails and water passages along the Hudson and Mohawk Rivers to the great lakes. Even airports later followed the pattern of urban concentration along the riverine link. As new transportation networks emerged, they displaced and in some cases replaced the previous ones. For example, in the center of the city of Schenectady, the Erie Canal, once the primary link to the interior of the continent, was filled in to become Erie Boulevard, and in the broader urban area an inter-city rail line that ran along the Mohawk River eventually became a recreational bike path. As occurred in many other American cities under the push from the petroleum and automotive industries, the intra-city trolley system was replaced entirely by buses and automobiles. The highways largely replaced passenger rail, but an Amtrak train continued to provide limited passenger service along the major cities of the old canal route.

The very capsulized history of a transportation field in a region of upstate New York makes it possible to outline some questions that can be asked in the study of the dynamics of technological fields:

1. Why were newer technological systems able to displace older ones? Did they offer significant advantages in terms of cost to consumers, profits to producers, or perceived technical efficiency (railroads for canals, cars for streetcars)? Did the changes involve significant conflicts among actors who had stakes in displacement? Did the social groups associated with losing technologies and industries, such as rail transportation, fight long rear-guard battles for their survival?

2. Why do some displaced technologies disappear completely (e.g., the local inter-city rail system between Schenectady and Troy and the streetcar system), whereas others shrink to small, specialty niches (such as the use of locks along the Mohawk River, where the Erie Canal still ran at the beginning of the twenty-first century but was primarily used

for recreational purposes)? What circumstances drive a sociotechnical system to extinction versus redefinition and niche specialization within a technological field?

3. How do technological distinctions coincide with class and other social distinctions? Which technologies are most and least accessible to the poor? How does the constant innovation associated with a technological field create a constantly changing politics of justice and access?

4. Under what circumstances are extinct or marginalized technologies revived for new purposes (such as an inter-city rail transit system, which is currently under consideration, albeit on a different route from the one that is now occupied by a bicycle path)? Do the new purposes reflect general historical transformations, such as increased population density, congestion, and air pollution?

5. How is one technological field linked to others, such as energy and recreation? Specifically, what is the link between energy (e.g., steam power) and transportation, or between recreation (e.g., bicycling and boating) and transportation? How do changes in one technological field (e.g., energy) disturb and reverberate across other fields?

6. What role did social movements, entrepreneurial firms, and user-oriented reformers play in the development of alternative systems, and how were their innovations incorporated and transformed into the dominant systems?

The concept of a technological field can help reveal lost technologies and envision the return of the repressed or displaced. Many of the alternative pathways discussed in chapters 5 and 6 (such as those in support of organic food, wind energy, recycling, and urban transit) draw on lost technologies as sources of inspiration for movements for technological innovation. The analysis of a technological field and its associated scientific research fields, which have a homologous structure of dominant and non-dominant networks, reveals what could have been and is still possible. It opens up the possibility of new design ideas that bounce off the history of what could have been. It teaches not only that the technical and political interact, but that they are brought into various alignments that have political effects and can be guided by democratic processes, or, at the minimum, agitated for by social movements.

The analysis of technology's history and design is simultaneously an analysis of its cultural meaning and social power. Power (which I define as the ability of an actor to mobilize resources to achieve a goal, which can include shaping agendas and goals of other actors) is embedded in

technologies in two major ways. First, in the commonsense and instru-
mentalist way technologies are objects of material culture that help
actors make changes in the social and/or material worlds. Technologies
have uses and impacts. Second, the design of a technology at any point
in time embeds a history of accepted and rejected design features. The
decisions that led to their selection, even when they were unintentional
or unconscious, have implications for the distribution of power.[9]

In order to understand technology and power in the broad sense that
I am outlining here, the concept of the sociotechnical network or system,
which played an important role in the conceptual development of technology studies, is situated inside the analysis of a technological field,
with its politics of selection among complementary and competing
sociotechnical systems or networks. As the literature recognized but did
not thoroughly conceptualize, there are power relations among networks
as well as between them and their allies or critics, and those power relations are encoded in design differences among technology, products, and
other aspects of material culture. The concept of the development of a
technological field is intended to serve as a way of situating system and
network analysis in the broader set of questions.[10]

Globalization and Object Conflicts

The analysis of technological fields is historical in a narrow sense, but it
is not yet situated within a broader historical sociology of modernity and
globalization. Whereas in science the historical circumstance of globalization has generated a significant literature on privatization and the
restructuring of the university and corporate research and development
laboratory, the control of technology development and diffusion by the
private sector is hardly a novel historical circumstance. Although universities are primary sites for developing the fundamental knowledge on
which invention and innovation rest, new technology, especially as new
products, is largely developed and produced within the private sector.
Consequently, the phenomenon of privatization that has occupied so
much attention in studies of science since the 1970s is of less interest in
technology studies. There is no news in the claim that much of technological development and diffusion is funded by and guided by the private
sector. Instead, the news associated with the era of globalization is the

increased attention to the universalization of technology production and design through standards and regulatory policy.

The regulation of material culture or technology is not new to an era of globalization; standards for buildings and infrastructure are an ancient phenomenon. However, several factors have converged to make the standardization of material culture increasingly salient in the world today. First, the standardization of technology is necessary in order for products to travel across diverse markets and government regulatory systems; firms need and want not only national standards but global harmonization. Second, because we have come to live in an increasingly technology-laden, denaturalized world, the product itself has become increasingly a manufactured rather than an extracted entity. As such, it embeds more of the intentionality of the producer and retains more of the reputation of the producer. The reputation that remains with the product as it travels through society motivates the producer to maintain brand quality and to institute standards that reduce liability. Third, the explosive growth of civil-society organizations and the differentiation within civil society of NGOs that call for the regulation of risky technology have created new awareness of the need for safety, health, and environmental standards. Finally, the regulatory function of government has undergone pluralization. Standards are set increasingly by international governmental organizations through harmonization treaties and by private-sector industrial associations. In turn, civil-society organizations negotiate directly with corporations to push them toward environmental, equity, and other standards for products, working conditions, and production processes; and corporations set standards that often privilege their market position.[11]

The growth of standards embraces not only production processes but also the design and use of objects: vehicles, buildings, food, drugs, biotechnologies, workplaces, databases, weapons, and a wide range of consumer products. Standards also govern increasingly regulated interactions between humans and animals, water, air, land, and ecosystems. Conflicts develop not only over the standards themselves but also over who has the power to set standards: local versus national governments, national governments versus regional states and trading blocks versus international organizations, or even governments versus private-industry-based organizations that develop voluntary industrial standards. Neoliberal and market-oriented ideologies may be marshaled to fight against some standards, but in general firms do not reject all types of standards and regulations. Rather, they embrace neoliberal ideology to undermine unwanted regulations that adversely affect profitability. Firms seek harmonization of regulations across political boundaries, just as they seek to transfer regulatory authority to their own self-governing, quasi-private standard-setting bodies. Firms require and want standards, albeit in the form of standards that they can control and limit, in order to utilize the products of other firms with confidence and to maintain consumer confidence in their own.[12]

The growth of standards that govern the production, use, consumption, and disposal of technologies and products suggests that capitalism under conditions of globalization has increasingly integrated the fundamental goal of production for profit with an emergent goal of production to standard. Just as modern rights guarantee, in principle, a minimal level of protection of the citizen from the state, so standards guarantee, in principle, a minimal level of quality for the product (a technology, a commodity, or even a service) with respect to the private-sector firm. The product itself has become redefined as an object that is the outcome not only of a production process but of definitional struggles that occur in various arenas over the standards and design of the object. I call the conflicts "object conflicts."[13]

Object conflicts are definitional struggles, simultaneously political, economic, and semiotic. The conflicts involve which objects should be released onto markets and, within categories of objects, which designs should be given priority over others. They involve governments, firms, individual consumers, and civil-society organizations, which interact in relationships of cooperation and conflict across various fields of action where the definitions of the proper object are worked out. Object conflicts have always existed in societies; however, in an increasingly technological world in which scientific expertise has been opened to the processes of epistemic modernization and products have become increasingly denaturalized and subject to standardization, they are becoming more salient.

Object conflicts can take the commonsense form of oppositional social movements that contest the appearance of an unwanted technological innovation and call for a moratorium on its development and diffusion, such as for nuclear energy, nuclear weapons, genetically modified food, chemical pollutants, and highways. However, there also can be object conflicts over the design and definition of alternative technologies and products for which there is broad support from the public or civil-society organizations, such as for renewable energy, organic food, remanufactured and reused goods, public transit, and green buildings. Object conflicts take place in various settings of interaction:

1. in the research setting over which scientific research programs will be funded and which will not, and, somewhat downstream, in the design setting where choices are made over which features of a product will be accepted for production and which will be left embedded in the shelved prototypes,

2. at the point of consumption over which products should be allowed on the market or widely offered to consumers (both individual and institutional), and over definitions of what constitutes a green or natural product,

3. in the regulatory and standards setting arenas over which technologies and products will be allowed to come on the market, which designs within an object category will be allowed, what production processes will be used to distribute the object, what standards will govern their use, and secondarily what processes will be used to govern regulatory standards and decisions,

and

4. over opposition that can emerge to the alternative technologies and products when they run into conflict with other goals, as occurs with wind farms.

Because object conflicts occur in diverse settings, the resolution of a conflict in one setting may have ramifications in others. For example, closure in a battle over standards may affect consumption decisions and research agendas, just as changes in research agendas and consensus shifts in scientific fields may affect consumption decisions and regulatory standards. Rather than focus on the design of objects as the stabilized outcome of a single controversy that leads to closure, I draw attention to the never-ending relations of conflict and cooperation over ongoing innovation in the design and construction of objects and their differential position in technological fields and markets.

Object Conflicts and the Environment

Of particular relevance for this book are object conflicts over environmental regulations and standards that would push industries to undergo ecological modernization. Comparative analyses have shown that regulatory push remains the crucial factor in motivating industries to undertake environmentally oriented changes in their production practices and product designs. For-profit firms tend to resist new environmental regulations and standards as expensive and unprofitable. Some firms have cited the high levels of environmental regulatory scrutiny in high-income countries as a primary factor behind their decision to relocate manufacturing to low-income countries with lax regulations. Because firms have historically viewed the relationship between environmentally oriented changes in production and profitability as a zero-sum tradeoff, they have often resisted changes and only made them when social-movement campaigns or high levels of scientific and public concern have pushed the government to do so.[14]

Although regulatory push (and behind it, social-movement activity and scientific research) has been and probably remains the main factor behind whatever trends toward the greening of industry can be found in most countries, there is considerable variation across industries and countries. For example, in many countries the chemical industry has undergone significant regulatory scrutiny, and it has been subjected to greater pressure to change industrial practices than most other industries. Likewise, pressures to undergo environmentally oriented reforms in production technologies and products have been much stronger in Western Europe than in the United States. In cases such as the Dutch chemical industry, as Arthur Mol has shown, partnerships among civil-society organizations, the state, and corporations have led to significant reforms. Although the reforms provide positive examples of what is possible in a favorable political situation, the reforms are in principle reversible; that is, they are subject to retrenchment as political climates change. Consequently, the role of civil-society pressure on industry, political parties, and government regulators is crucial to maintain and deepen the greening of industry, and to initiate it where it is absent.[15]

In contrast with the zero-sum frame of profits versus the environment that neoliberal opponents of environmental regulation tend to utilize, a second strategy for technological change challenges the assumption of a tradeoff. A new industry of consultants, university researchers, and other specialists provides expertise for firms that are willing to view environmentally oriented changes as technological innovations in the form of investments that bring direct financial returns as well as indirect returns to brand and corporate image. For example, by closing the loops of manufacturing waste, firms can reduce waste disposal costs and capture profits from waste reuse. By emphasizing the positive-sum politics that link the profitability goals of the firm and the environmental goals of social movements and sometimes the state, the eco-innovation frame can be quite appealing to both firms and environmental groups.[16]

However, in practice the eco-innovation approach opens the door to the second type of object conflicts, ones that emerge in the design of alternative production practices and products. For example, a furniture factory may invest in new cogeneration technology that allows it to divert wood shavings from a landfill by recycling them on site for heating and electricity generation via burning. The eco-innovation may appear to generate profits while reducing the firm's environmental impact, but the change may result in a net increase in greenhouse-gas emissions in contrast with landfilling, and it could also result in higher levels of particulate matter in the air of the neighborhood around the plant. As a result the design innovation, the cogeneration plant, could itself become the site for additional object conflicts in the form of how to regulate distributed energy generation based on wood burning, or what kinds of emissions technologies are appropriate for industrial wood burning.

In general, the existence of object conflicts, particularly those between grassroots groups and large business organizations, points to the politicization of technological innovation and design that is parallel to the politicization of scientific research agendas discussed in the previous chapter. One can find examples of such politicization in the past, such as the concern of working class and labor organizations with the effects of mechanization on the quality of work and availability of jobs. There are also instances of working class campaigns that took machines out of the black box and entered into the politics of design. But since the middle of the twentieth century, the politics of new technologies have increasingly come to focus on design. The new focus of the politics of technologies draws attention to standards not just of production processes in the workplace but also of products that circulate among business and home consumers. Regulatory bodies, standards organizations, and watchdog groups have proliferated to define best practices and acceptable standards, and movements have recruited experts who can help articulate an effective politics of opposition that focuses on more and less desirable design features. Increasingly, the politics of technology in a period of globalization is less about whether or not a new technology should be introduced, although that form of politics has not disappeared, and more about the design choices that should be made before the introduction of a new technology. For example, public interest organizations may be against genetically modified food if it avoids regulatory scrutiny, forces farmers into new relationships of dependency, exposes ecosystems to unknown damages, and exposes small segments of human populations to allergic reactions, but other designs of genetically modified food, such as public-sector projects that help to develop forms of rice that have higher levels of disease-reducing vitamins, may be welcomed by the same organizations.[17]

Alternative Pathways and Industrial Innovation

In chapters 4 and 5, I will examine alternative pathways for social change that are particularly relevant to the object conflicts American society faces in regard to environmental issues in the era of globalization. The next two chapters will examine alternative pathways in the United States that seek changes in industry and technology. (See table 3.1.) Often the activists and advocates of alternative pathways draw on the counter-expertise of scientists who are outside the mainstream of their fields, and they seek a shift in funding priorities to correct what they perceive to be cases of undone science. There are two major types of industry-oriented social-change action: industrial opposition movements (IOMs) and related social-change action, which focus on the remediation of environmental risks and problems; and technology- and product-oriented movements (TPMs) and related social-change action, which focus on innovation of design processes.[18]

Although there is a small literature on the topic, my focus of attention on the role of social movements and reform movements, activists and advocates, and related social-change agents in industrial innovation may

Table 3.1
Alternative pathways in industry and technology.

Fields of action	Industrial opposition movements	Technology- and product-oriented movements
Food and agriculture	Pesticides, GMO food, factory farms	Organic
Energy	Nuclear energy, fossil fuels	Renewable energy
Waste and manufacturing	Pollution, local toxic exposure	Recycling, zero waste
Infrastructure	Highways, sprawl	New urbanism, smart growth, green building
Finance	Environmentally damaging investments	Responsible investing and consumption

seem counterintuitive. The literature on social movements tends not to address the topic except through its attention to the ways in which IOMs, such as the anti-nuclear-energy movement, have contributed to the cessation of a particular type of technology. However, my argument, which builds on and develops that of Andrew Jamison, is that social movements have a generative capacity as well. They not only oppose certain types of new technologies and products but they help to develop and diffuse alternatives. In an era of globalization and market-oriented government policies, social movements have helped to politicize consumption and, in the process, to develop new markets and industries. They have also pointed to ways to design sociotechnical systems so that industrial innovation is linked to community control and social justice.[19]

4

Industrial Opposition Movements

Industrial opposition movements (IOMs) aim to stop a particular technology and often the broader system of relationships in which it is embedded, including production practices, industrial and consumer products, pollution generated in production and consumption, firms that use and diffuse the technology and product, associated research agendas in both firms and universities, and state and industrial policies that govern the network. IOMs may focus on stopping research and development for a proposed technology, halting implementation of a technology that is developed and ready for market, and/or resisting the diffusion of a technology that is already on the market. All the strategies are aspects of the IOM's overarching goal of achieving a moratorium, that is, an end to a particular technology and production practice. In turn, industry may respond by suppressing and ignoring the activists, halting production of the targeted technological system, redesigning it or innovating by developing an alternative technology, or some combination of all three. By leveraging a response from the targeted industry, IOMs can affect industrial innovation, and they can shift the direction of the history of a technological field.

Organizationally, the IOM generally consists of a network of social-movement organizations that coalesce around specific campaigns and may be coordinated through umbrella organizations. Although the IOM generally approximates, more than the other types of alternative pathways considered here, what the literature readily recognizes as a social movement, in some cases the mobilization remains at a modest level of scope that could be characterized as only activist networks and campaigns. Professional reform movements of scientists, other professionals, and entrepreneurs may also coexist with the IOMs, and the IOM organizations will sometimes attempt to recruit experts from scientific and technical professions to provide needed counter-expertise. There is also likely to be interaction with the other types of pathways. Although the moratorium is the primary goal, the IOM may have secondary goals of support for alternative technologies, democratic or local control of technological decision making, and access for the poor. Consequently, there is both interchange and overlap with other types of pathways, not only in terms of goals but also in terms of life histories and organizational relationships.

As theoretical categories, industrial opposition movements and technology- and product-oriented movements (TPMs) are distinguished by their different goals, but in practice the organizations and movements are often closely linked. Some organizations have both IOM and TPM missions, and individuals often have biographical trajectories of work in both IOMs and TPMs. IOMs need TPMs to legitimate the call for the moratorium; in turn, the TPM supports the alternative that can replace the technology that should be phased out. Without the TPM the IOM would be vulnerable to the criticism of mere negativism. Likewise, the TPM relies on the work of the IOM to draw attention to the shortcomings of existing or new industrial technologies, the benefits of the alternative, and the need for research funding and consumer support of the alternative.

The incorporation of IOM goals by political and industrial elites tends to occur through the implementation of a partial moratorium. Elites may put an end to some of the most egregiously offending technologies and products, or they may halt new production of a technology or product, but they generally do not grind an entire industry to a halt. The transformation of grassroots goals for a full moratorium into the partial moratorium may coincide with the polarization of the movement and the typical divisions that occur as elites both recognize and co-opt a movement's goals for change. However, the patterns, extent, and waves of mobilization vary considerably across the fields of action.[1]

The sections that follow will discuss the ways in which IOMs have contributed to complete or partial moratoria for technological systems in five main industrial fields: food and agriculture, energy, waste and manufacturing, infrastructure, and finance. I have also been able to identify similar patterns in other industrial fields, such as medicine and the media, but the other fields are not considered here because they are less directly related to the environmental focus of this book. Two limitations of the discussion, both in the present chapter and in the two chapters that follow, should be emphasized. First, I have focused on a specific issue in the history of alternative pathways: the incorporation and transformation process, that is, the pattern of absorption of alternative pathway goals into mainstream industries and government policies, and the transformation of the design of technologies and products that have occurred in the process. In other words, I am focusing on the partial success or failure of the pathways. My goal is not to become immersed in the rich particulars of each history but instead to provide enough information to understand the historical trend for each pathway with respect to the research question. Second, the work by social scientists and historians is very uneven; I have had to face my own problems of undone social science, ethnography, and history. In some cases, such as the anti-nuclear-energy movement, the peer reviewed literature is well developed, and the incorporation and transformation process can be examined in detail with confidence. In other cases there is some peer reviewed professional literature on the history and sociology of the alternative pathway, but it is very limited. In yet other cases I have had to assemble the history from primary sources, such as organizational histories from websites and industry publications. As a result there is an inevitable unevenness to the analysis, and questions that can be asked and answered in some detail for some pathways can only be examined in a preliminary form in others. If anything, my analysis will suggest many future studies that could benefit both the agents of change and the broader research fields that study social movements, social and technological change, and the environment.

Food and Agriculture

In 1900 the United States was a largely rural society, but the rural/urban proportions had reversed by the end of the twentieth century, so that in 1990 about 75 percent of the population lived in urban areas. The number of farms declined from about 6 million at the beginning of the twentieth century to under 2 million at the century's end, and farm employment dropped from 11 million to about 3 million. As the number of farms and farmers declined, farm size increased from an average of 150 acres to more than 400 acres. Innovations in agricultural technologies aided the scale expansion of farm size and the industrialization of agriculture, but the changes also set the stage for industrial opposition. Pesticide contamination of farm workers and ecosystems, depletion of aquifers, contamination of groundwater, salinization of soil, ecological risks associated with genetically modified food crops, and waste from concentrated animal feeding operations (CAFOs) were among the concerns that underlay the search for what became known as "sustainable agriculture." Likewise, the industrialization of inputs and the growth of processed food coincided with a new range of food-related health concerns, including pesticides in food, overuse of antibiotics, transmission of animal diseases to humans, loss of nutrition due to food processing and poor soil conditions, allergic reactions to genetically modified food, and health effects of food additives.[2]

During the late twentieth century there was no single, dominant IOM in the food and agricultural field, but there were various campaigns and mobilizations on environmental issues such as pesticides, genetically modified food, and CAFOs. Anti-pesticide mobilizations were able to align health and environmental concerns, and they achieved some victories. One example is the prior informed consent principle, which requires exporting countries to obtain consent from importing countries before shipping hazardous substances. By the late 1990s several of the worst pesticides in food had been banned, and the principle of prior informed consent had been accepted in international law. However, the anti-pesticide organizations achieved only a partial moratorium by obtaining a ban on some of the most hazardous chemicals and better regulations over pesticide use. After decades of work they had not achieved a fundamental change in an agricultural production system that relied heavily on pesticide control of large monocrops.[3]

As the anti-pesticide network developed, its mission diversified to include campaigns against genetically modified (GM) food. The anti-pesticide groups were only one of many types of organizations that have formed coalitions to oppose GM food. The anti-GM-food campaigns were less well developed in the United States than in Europe and South Asia, but American organizations nevertheless formed substantial coalitions that advocated banning, reviewing, or, at the minimum, labeling such food. An analysis of the organizational membership in the Turning Point Project, a series of advertisements placed in the *New York Times*, revealed a broad coalition of organizations. About three-fourths of the organizations were food, agricultural, or environmental NGOs, and the rest were science based, consumer, animal rights, and left/labor. Street protests also took place in the United States, such as in June 2003, when thousands gathered in Sacramento, California, to protest a conference of agriculture officials on GM food. In Kentucky small tobacco farmers also mobilized to head off a state government plan to invest tobacco settlement funds in biotechnology. In that state the small farmer coalition brought about a shift in funding to sustainable, local agricultural development, a change in policy that also suggests the close connection between IOM and TPM goals that can sometimes emerge. In general, the anti-GM-food protests and lobbying in the United States did not achieve the breadth and scope associated with some of the other, late-twentieth-century American social movements, and reform legislation at the national level introduced in 2002 was not successful.[4]

Again, the pattern of a partial moratorium, arguably more limited than the one achieved by the anti-pesticide organizations, characterizes the fate of the anti-GM-food movement to date in the United States. Once industry agreed to restrict the more dangerous genetic modifications, such as genes that could trigger allergy and death in a small percentage of the population, production has moved forward for some of the largest crops. In the United States the movement has not been able to achieve even the limited goal of labels for GM food in consumer packaging. The industry rejected labeling because it believed that it would imply that GM products are unsafe. Nevertheless, some retail food corporations have gradually distanced themselves from GM food, and in 2004 Monsanto announced that it would abandon field tests of GM Roundup Ready wheat. Greenpeace claimed victory for the instance of partial moratorium.[5]

In addition to anti-pesticide and anti-GM-food campaigns, a third type of IOM in the food and agricultural field is opposition to factory farms or CAFOs. Coalitions of environmental and health organizations have drawn attention to concerns ranging from watershed pollution to antibiotic overuse and the risks of mad cow disease and avian flu. State medical societies and the American Public Health Organization have joined with environmental groups such as Friends of the Earth, the Union of Concerned Scientists, Public Citizen, the Sierra Club, the Citizens Environmental Coalition, and the Grace Factory Farm Project to push for a moratorium on factory farming and the overuse of antibiotics in animal livestock. Although the European Union decided in 1998 to ban some antibiotics in livestock, the U.S. Department of Agriculture opted instead for more research. However, researchers who have documented the health and environmental risks of CAFOs have encountered significant suppression, especially when they are located in public universities. Because of the influence of CAFOs on state legislators, in some cases the state government has exerted pressure on researchers to abandon their work. Notwithstanding the resistance that researchers and environmental and health organizations have encountered, some small gains have been made. For example, after years of advocacy, a coalition of environmental and health organizations finally convinced the Food and Drug Administration to ban one type of antibiotic used as a growth promoter for chickens.[6]

Although attempts to implement a moratorium on antibiotics to date have encountered only limited success at the national level, another strategy of moratorium has emerged the local level. When faced with the negative local environmental effects of CAFOs and public outrage at the grassroots level, some county governments have either banned factory farms or restricted their practices. Efforts are particularly advanced in Pennsylvania, where the Community Environmental Legal Defense Fund has been actively helping local governments. In the 10 years after its founding in 1995, the organization helped a dozen communities to ban factory farms and more than 40 communities to end the dumping of sewerage sludge onto farmland. Here, the moratorium is complete, but it is highly localized and sometimes has been reversed after subsequent litigation by CAFOs. The companies have also attempted to supercede local laws by influencing the state legislature and governor's office. As conflicts with the state legislature and animal feeding companies have heated up, local city councils have increasingly raised the stakes. Ultimately, their frustration with the way that the courts and state legislature have protected corporate rights over citizen rights has led some counties to shift their strategy to a direct legal challenge to the idea of corporate personhood.[7]

In summary, the three IOMs (those opposing pesticides, genetically modified food, and factory farming) all have called for a moratorium on selected technologies and production processes associated with industrial agriculture. The first two IOMs have operated more at a national and international level, and they can claim limited success. In the case of pesticides, the successes have been in the political arena, whereas for GM food the successes have been more with consumer corporations that are concerned with brand dilution and have opted not to purchase GM food. In contrast, opposition to factory farming has not yet translated into major policy changes at the national level other than the ban on a type of animal feed antibiotic. To date, bans on factory farming have occurred more at the local level, where the health and environmental risks are concentrated enough to generate the kinds of powerful mobilizations that are continuous with environmental justice struggles.[8]

Why have the anti-pesticide groups achieved success, particularly at the international level? One hypothesis is that risks must be widespread and alarming to the public, such as those posed by some pesticides, and the risks must be documented through scientific research, which requires that the research agendas in fields such as public health and toxicology do not relegate such research problems to undone science. If the risks are substantial, documented, and publicized, environmental organizations can use alignment with the health concerns of individuals to leverage popular support in favor of regulatory changes from governments. At an international level the frame of dumping and neocolonialism can be brought together with the health and environmental concerns into a more powerful bundle of motivating factors. Where foreign governments are independent from the pressures of multinational pesticide and chemical companies, there is a political opportunity to convince the governments to take stands against pesticide dumping. Likewise, genetically modified food is seen as one more example of the dominance of American corporations over local foodways, and nationalist or anti-American frames can also be combined with claims of health and environmental risk to motivate mobilizations against GM food. Even where governmental strategies do not work out completely, the environmental organizations can leverage public perception of risk and anti-imperialist sentiment to get brand-conscious consumer corporations (such as food-processing companies) to impose their own moratoria rather than risk loss of market share.[9]

One reasonable hypothesis is that where health risks are localized, as in factory farming, or not well documented, as in many forms of GM food, then the general alignment with health risks is missing, and consequently the federal government's response will be weaker. At the local level, where direct health and environmental risks are widespread and highly visible, the case of factory farming shows that local government response can be quite strong, but local governments may find themselves outgunned by large corporations. Here, outrage based on a rights frame—that is, the rights of local government to self-determination over their land use and zoning policies—can help fan a second wave of mobilization after the failure of local ordinances based on health and quality of life considerations. The shift from a frame of technological design to political justice and local sovereignty links the IOM with the localist pathways discussed in chapter 6.

One can see that IOMs have had a variable and, in the United States, fairly limited influence on the general direction of technological fields in industrial agriculture. Even where they have achieved a moratorium on specific aspects of industrial agriculture (such as specific pesticide bans), the moratoria take place in a context of continued innovation from the chemical and biotechnology industries. As a result the environmental organizations are on a treadmill of activism. A more substantial influence on the technological fields would occur if there were moratoria on broad categories of technologies, such as no antibiotics in animal feed or no chlorinated chemicals in agricultural production. Such changes would probably require a modernization of the regulatory approval process to open it up to greater public participation. However, because the battles for more general moratoria and improved public participation in regulatory decision making are difficult to win, the organizations end up fighting a series of valiant but endless rear-guard battles for remediation.

Energy

As with agriculture, the energy industry has also undergone continuing consolidation, but in contrast with the agricultural pattern of the replacement of small, family-run farms with large farms, in the energy industry consolidation has involved mergers among large corporations.

In contrast with agricultural policy, which has had the image and place
of the small family farm as a point of reference, when policy makers
began to develop responses to air pollution and dependence on foreign
oil during the 1960s and the 1970s, the policies were articulated within
a framework of a corporate-controlled industry. Centralized, grid-
controlled energy met the goals of large energy corporations as well as
the defense industry, and nuclear energy emerged as a boundary object
that met the needs of both. However, as with the growth of CAFOs, the
choice of nuclear energy set industrial and policy elites on a collision
course with communities located near the sites, and the stage was set
for the growth of the anti-nuclear-energy movement. Because the literature on the anti-nuclear-energy movement is more developed and the
movement has been the dominant IOM in the energy field, at least in the
United States during the late twentieth century, its history can be considered in some detail in this section.

Organized opposition to nuclear energy in the United States began
much earlier than the 1970s, when the movement achieved widespread
media attention. In the 1950s Democrats in Congress and the United
Auto Workers joined to stop plans for the Enrico Fermi reactor in
Detroit, and the Sierra Club joined with residents of Bodega Bay,
California, to stop a planned nuclear energy power plant. The Bodega
Bay campaign divided the Sierra Club and marked the beginning of the
organization's long shift from conservationist politics toward the
hazards and risks of industrial technologies. Friends of the Earth split off
from the Sierra Club in 1969 and went on to provide anti-nuclear leadership, but during the 1970s the post-split Sierra Club also became
increasingly anti-nuclear. Divisions within the scientific community also
provided another important impetus to the development of anti-nuclear
movement. For example, a controversy developed over the safety of
allowable radiation doses, and John Gofman, a prominent scientist who
concluded that allowable dose thresholds were too high, was forced out
of Lawrence Livermore Labs. In 1969 the newly founded Union of
Concern Scientists provided counter-expertise on nuclear energy, and in
1971 Gofman founded the Committee for Nuclear Responsibility.[10]

During the early 1970s new orders for nuclear power plants acceler-
ated, and the various organizations and campaigns coalesced into an
IOM. One major coalition organization, the Consolidated National

Intervenors, included 60 groups and advocated the design of safer reactors rather than an all-out freeze. Major NGOs included the Nader group Critical Mass, Friends of the Earth, the Sierra Club, and the Natural Resources Defense Council. Significant organizational strength for local action came from a goal shift of 1960s anti-war organizations, such as Mothers for Peace, which in 1973 expanded its activism into nuclear energy. Likewise, the New Age organization Creative Initiative Foundation figured prominently in a 1974 campaign in California.[11]

By the mid 1970s, opposition to nuclear energy had shifted away from a focus on nuclear energy safety, an approach that brought attention to the technical issues of reactor design, toward a more clearly anti-nuclear-energy position and the politics of the complete moratorium. At the same time the movement's organizations and repertoires underwent differentiation. The 1977 protests and arrests of the Clamshell Alliance at the Seabrook site in New Hampshire drew media attention and emulation, as did similar protests against the Diablo Canyon plant on the West Coast led by the Abalone Alliance. Those and other anti-nuclear alliance organizations (such as the Catfish, Crabshell, and Cactus alliances) practiced a direct-action repertoire and drew support from residents who were located in close proximity to nuclear reactors or planned reactors. Consequently, the mobilizations had a "not in my back yard" (NIMBY) flavor similar to the subsequent anti-CAFO and anti-toxics movements.[12]

In addition to civil disobedience at existing and proposed nuclear reactor sites, the movement carried out campaigns of state-level ballot initiatives, lobbying, and appearances at siting hearings. One influential development was the elimination of construction work in progress laws, which began in 1976 and spread rapidly to many states. The laws had allowed utilities to finance construction by placing a charge on customers' utility bills, and the elimination of the laws, together with high interest rates, significantly increased the cost of nuclear reactors. In 1975 there were 21 new orders for nuclear power plants; in 1976 there were no new orders and 17 cancellations. The red light on the economic outlook for the nuclear energy industry coincided with the rising tide of grassroots protests against nuclear power plants and the increasing divisions among experts and professionals. Although most analysts argue that the changed economic picture, rather than the protests, was the more decisive factor in the demise of new nuclear energy construction, the two factors were not independent, because anti-nuclear activism and campaigns had led to legislative changes that increased costs.[13]

The year 1976 also represented a turning point in terms of the incorporation and transformation of anti-nuclear sentiment into public policy. In that year Jerry Brown, who served as California's governor from 1974 to 1982, began his shift to an anti-nuclear position that culminated in his rejection of the Sundesert plant in 1978. His presidential bid in 1980 included a call for a total moratorium on nuclear energy and brought the issue into the presidential race. Similarly, in 1977 President Carter opted not to support plutonium as a fuel source, and a year later he developed a national energy plan that reflected some of the concerns raised by the anti-nuclear and broader environmental movement. The plan split environmental groups between those that supported Carter's pro-conservation focus and those that opposed his acceptance of modified nuclear energy based on greater safety. In 1979 the crisis at the Three Mile Island nuclear facility in Pennsylvania, together with the release of the popular movie "The China Syndrome," raised public fears and contributed to the turn of general public opinion against nuclear energy. During the 1980s the remains of the anti-nuclear-energy movement became increasingly local in orientation, and activist networks focused on specific policy issues, such as waste management.[14]

After the anti-nuclear-energy movement could claim partial success in the form of a halt to the construction of new nuclear energy plants, the organizations and leaders shifted to opposition to nuclear weapons. Although generally considered a peace movement rather than an environmental movement, the historical connections with the anti-nuclear-energy movement have often been missed, and one might remember that the environmental risk posed by the prospect of nuclear winter makes global warming pale in comparison. The transition from energy to weapons has been dated to 1977, when the Union of Concerned Scientists shifted its goals and the Clamshell Alliance, which linked the two kinds of anti-nuclear struggle, was founded. Although some anti-nuclear-energy organizations, such as Critical Mass, rejected the linkage, the direct-action organizations embraced it. One significant anti-weapons group, the Livermore Action Group, began under the umbrella of the Abalone Alliance, which itself did not survive the goal shift to anti-nuclear-weapons activism. As the anti-nuclear-weapons organizations developed, some rejected the linkage to the anti-nuclear-energy movement. For example, the Nuclear Weapons Freeze Campaign organization dissociated itself from the energy issue, as did Physicians for Social Responsibility, which grew rapidly during the early 1980s and forced board member John Gofman to leave.[15]

Although the anti-nuclear-energy movement dwindled during the 1980s, other forms of oppositional energy activism and campaigns emerged. In the 1980s and the 1990s environmental organizations shifted to concern with the atmospheric and health effects of fossil-fuel pollution. Friends of the Earth, Greenpeace, and the Sierra Club developed litigation directed at the fossil-fuel policies of the federal government and international organizations. The Natural Resources Defense Council, American Lung Association, and Union of Concerned Scientists assisted in campaigns to reduce diesel emissions, which were linked to respiratory illnesses in children who rode school buses and in residents of areas close to bus lines and bus yards. Public Citizen's Critical Mass and the Union of Concerned Scientists developed programs and information campaigns in support of "clean energy." The organizations called for the restructuring of the energy industry and the development of renewable energy portfolio standards (that is, a minimum percentage of total energy consumption from renewable energy). Although several of the organizations continued to be active on issues of nuclear energy, especially the politics of nuclear waste storage, the focus shifted away from a moratorium directed at a specific target. Instead, the organizations advocated policies that would mitigate global warming and convert energy production from fossil fuels to renewables. At this point, opposition action that targeted fossil fuels was increasingly interwoven with the TPMs that developed to support solar, wind, and other renewable energies. Unfortunately, the concern with reducing greenhouse-gas emissions has created new divisions within the environmental movement, because some environmentalists have sided with the nuclear energy industry's call for developing new nuclear power plants based on their potential to mitigate greenhouse-gas generation, whereas others, such as Public Citizen, have begun campaigns against what they term "nuclear relapse."[16]

Although many environmental organizations have campaigns in favor of clean or renewable energy, the organizations have not mobilized as much public support as was mobilized by the anti-nuclear-energy movement of the 1970s. One case where oppositional politics have been successful in the energy field since the 1990s is the banning of methyl tertiary butyl ether (MTBE), a fuel additive that was used to meet the Clean Air Act Amendments of 1990 but was found to leak from gasoline tanks and pipes into groundwater supplies. By 2005, nearly half of the states had enacted some type of ban on MTBE, and in 2006 a shift in the Environmental Protection Agency policy on oxygen content facilitated the phase-out of MTBE. Again, the politics of the moratorium are accompanied by developments of an alternative. Although the federal government policy shift did not stipulate a shift to ethanol as a mandatory replacement, the ethanol industry benefited.[17]

In summary, the pattern of the moratorium in the energy field, even for MBTE, has been partial. Although the anti-nuclear-energy movement shifted from a design-and-safety-oriented approach to a complete moratorium, it only achieved a moratorium on new construction. As in the agricultural field, the politics of the moratorium have relied on the knowledge claim that risks are imminent (such as the threat of a nuclear reactor meltdown, high levels of air pollution, or drinking water contamination from MTBE). Furthermore, the long-term, sustained mobilization involved a local component, because local communities felt threatened by new power plants (just as they felt threatened by incinerators and factory farms). Another factor that contributed to success was the emergence of a countervailing industry available to support the goal of reduced use of a technology or phase-out, such as the emerging wind and ethanol industries. Achieving even the more modest goal of a partial moratorium on an entire industry also involved indirect strategies, such as cutting funding mechanisms for new construction, rather than outright legislative victories. Still, the principle IOM in the energy field could claim to have played a significant role in the shift in the field of energy technologies through the 30-year moratorium on new nuclear energy power plants, which was not challenged until the early 2000s.

Waste and Manufacturing

The IOMs that I have classified under the field of waste and manufacturing are extensive, because they embrace the activity of the mainstream environmental movement and the environmental justice movement. There is considerable material written on their history, so the challenge in this section is how to extract from the history enough detail to be able to chart out the patterns of incorporation and transformation and the shifts in the technological field in the United States since middle of the twentieth century. One convenient historical starting point is the publication of Rachel Carson's *Silent Spring* in 1962, which encouraged public concern over radioactive fallout, air pollution, and water pollution. Under the Democratic administrations of Presidents Kennedy and Johnson, various environmental laws were enacted, among them the Clean Air Act of 1963. During the administration of President Nixon addition environmental legislation was passed, including the National Environmental Policy Act (which established the Environmental Protection Agency), the Clean Air Act of 1970, and the Clean Water Act of 1972. Environmental legislation continued to be enacted throughout the 1970s, including the Safe Water Drinking Act; Federal Insecticide, Fungal, and Rodenticide Act; Resource Conservation and Recovery Act (for hazardous and solid waste); Toxic Substances Control Act; and Comprehensive Environmental Response, Compensation, and Liability Act (Superfund).[18]

The existing work on the history of the environmental legislation of the 1960s and the 1970s suggests that the environmental movement's role in bringing about the legislation was dubious. With some important exceptions, prior to the late 1960s the movement was largely oriented toward wilderness and wildlife preservation issues rather than industrial pollution. Some of the older preservation and conservation organizations underwent changes, sometimes in ways that linked concern with industrial pollutants to preservationist issues. One example is the Sierra Club's goal shift, discussed above, and another is the Audubon Society's concern with the effects of industrial pollution on birds. In addition, during the 1960s and especially in the wake of the first Earth Day, which took place in 1970, a new group of organizations emerged to focus more explicitly on industrial pollution.

Regarding the issue of incorporation and transformation, the pattern during this period is rather different from some of the other IOMs concerned with environmental issues. Although the mainstream environmental organizations, including the groups founded by Ralph Nader, played a role in supporting some of the environmental laws of the 1960s and the 1970s, it is not accurate to claim that a grassroots social movement mobilized against pollution and forced a change on Congress and the White House. There was no buildup of a mass social movement against industrial pollution that resulted in a response by political elites, who incorporated the calls for moratoria on pollution and transformed them into limited policy remedies. Rather, Democratic Party leaders responded more directly to public concern and scientific studies, and party leaders such as Senator Gaylord Nelson provided leadership in mobilizing public support for the first Earth Day rallies of 1970s. Republican Party leaders such as President Nixon also embraced the environmental issue, but more due to political calculation based on reelection concern, especially the fear that he might face the pro-environment Senator Edmund Muskie in the 1972 election. Environmental issues also played a significant role in the election of President Carter in 1976, and he moved environmentalists into some influential administrative positions. The broader point is that during this period support for environmental legislation to begin to restrict pollution was widespread, and elected public officials were responding to public concern and scientific research. Unlike in some of the other fields, in this case IOM mobilization was to some degree short-circuited by the direct response from Washington.[19]

The legislative response of the 1960s and the 1970s, together with the continuities between the preservationist movement of the early twentieth century, account for the well-known inside-the-beltway orientation and moderate tactical repertoire of the side of the late-twentieth-century environmental movement that was focused on opposition to industrial pollution. Oppositional politics within mainstream environmentalism tended to take the form of providing expertise and support for legislative reform. Some of the new groups, such as the Environmental Defense Fund and the Natural Resources Defense Fund, also used litigation to enforce laws that were already on the books. The tactic of mass mobilization and protest passed to the anti-nuclear movement, the environmental justice movement, and various other wings of the broader environmental movement.[20]

By the 1980s, the waves of legislation and litigation of the 1960s and the 1970s had resulted in a backlash by the private sector, which rebelled against both the cost of compliance and the use of non-voluntary, end-of-pipe metrics. After the election of Ronald Reagan in 1980, the mainstream organizations formed a defensive coalition that became known as the Group of 10. Although the memberships of the organizations soared in response to threats to the environment, the coalition was unable to develop a coherent agenda and instead produced a weak joint statement that focused on population control. Aggressive litigation became increasingly difficult as Republican appointments to the courts increased. By 1990 public policy had shifted away from command-and-control legislation toward market mechanisms such as emissions trading. Some of the nationally oriented, mainstream environmental organizations and leaders played a role in developing and supporting the market-based approaches. Often seen as co-opted by neoliberal ideologies and partnerships with large industrial corporations, I would suggest instead that the shift to market-based policy remedies was an example of how the goals of pollution remediation were incorporated and transformed by industrial and political elites that had regrouped and reformulated after the 1970s. The change did not mean that the mainstream environmental organizations ceased to play an oppositional role. Many of the leading environmental organizations continued to support policy changes aimed at reducing water and air pollution, global warming, pesticides, and toxic exposure, and from the 1980s on they often fought rear-guard battles to protect environmental legislation from rollbacks.[21]

In an ideal world, social movements would not be necessary; elected officials would protect the environment and public health by funding areas of undone science and using the research to make policies of general social and environmental benefit. To some extent, at least more so than in the other IOMs described in this chapter, the legislative history of the 1960s and the 1970s approximates a government that was responding to scientific knowledge and public concern about health and environmental risk. Although the response was generally inadequate, often sidetracked, and by the 1980s increasingly reversed, the mainstream environmental organizations did seek to remedy some of the most pressing environmental problems. In such a situation the environmental organizations that were concerned with national policies utilized lobbying and media campaigns that were more consistent with a political reform movement or interest group politics than a protest-oriented social movement. In other words, the incorporation of mainstream environmental organizations was more subtle, because many of them already had an insider status. To the extent that one can identify an incorporation and transformation of their agendas, it is in the shift from mandatory pollution controls and litigation strategies of the policies of the 1960s and the 1970s to the market-oriented, cap-and-trade policies of the 1980s and after. As the mainstream environmental organizations increasingly worked in partnership with industry and regulatory agencies, mobilization that utilized protest against waste, pollution, and toxic exposure shifted increasingly to grassroots groups in the environmental justice and anti-toxics movements.

The literature on the environmental justice movement is also extensive, and the brief discussion here will extract from that literature an analysis of the multi-decade pattern of incorporation and transformation, as well as shifts in the technological field. Historically, the environmental justice movement drew significantly on the civil rights experience of African-Americans and other ethnic minority communities. The emphasis on rights of access to clean air, water, and land in the environmental justice movement suggests continuities with action that I will later discuss as access pathways. Formative events in the history of the environmental justice movement in the United States include the drowning of an African-American girl in a Houston garbage dump in 1967 and plans to set up a polychlorinated biphenyl (PCB) waste site in Warren County, North Carolina, in 1982. In addition to grassroots mobilizations by the communities, the environmental justice movement grew from the contribution of other streams. For example, health-oriented organizations emerged to study and reduce exposure to environmental risk factors for disease, and an academic reform movement of scholars (generally marginalized within the scientific fields in which they were located) established environmental justice research centers and developed crucial research in support of the epistemic claims of grassroots groups. There were also parallel struggles by Native Americans against the siting of toxic waste sites, extractive projects, and other sources of pollution on reservations; and labor unions fought toxic exposure in the workplace, including the pesticide exposure of farm workers. In addition, some of the mainstream or second-wave environmental organizations eventually diversified their missions to include environmental justice issues.[22]

Yet another stream that is associated with the environmental justice movement, and sometimes distinguished from it, is the anti-toxics movement. That IOM has another history of formative events, with the Love Canal tragedy in the late 1970s often mentioned as the point of origin. The Love Canal mobilization involved a middle-class neighborhood that suffered health effects from unattended waste and failed waste remediation. The social address of the anti-toxics movement is more variable, but it includes white neighborhoods, often with women in leadership positions, and often with much less social-movement organizing experience than African-Americans in the environmental justice movement, some of whom have a background in the civil rights movement. The anti-toxics movement has claimed as one success the development of Superfund legislation, which provided some support for cleanup of brownfields. Both the environmental justice and anti-toxics movements can also point to significant growth figures. For example, the Citizens Clearinghouse for Hazardous Wastes grew from 600 affiliated community groups in 1984 to 5,000 in 1988 to 10,000 in 1995.[23]

As with some of the IOMs in other fields, the environmental justice movement draws its strength from the direct health threat that industrial waste and pollution pose to households, children, and communities. Notwithstanding the NIMBY basis for mobilization, the movement has also developed a national and international political presence. The shift toward a national strategy and then back away from it may have coincided with the opening of political opportunities during the administration of President Clinton, although there was growing mobilization during the Reagan-Bush years. During the 1990s there was an incorporation of environmental justice goals into national policy. After the 1991 First National People of Color Environmental Leadership Summit, the Environmental Protection Agency responded by creating an Environmental Justice Office. Likewise, Congress considered, but did not pass, an environmental justice bill. In 1993 the Environmental Protection Agency established the National Environmental Justice Advisory Council, and a year later President Clinton issued Executive Order 12898, which directed federal agencies to address public health and environmental inequalities. Although well intended, the federal response did not put an end to the ongoing targeting of low-income and ethnic minority communities for a disproportionate share of the environmental toxic burden.[24]

The rather weak response during the Clinton years did not improve during the subsequent political administration. A 2001 restatement of the agency's environmental justice agenda did not emphasize low-income and ethnic minority populations, and in 2004 a report from the Environmental Protection Agency's Office of the Inspector General claimed that Executive Order 12898 had not been fully implemented. Regarding environmental justice policy during the presidency of George W. Bush, one could credibly make the case that the policy response at the national level largely transformed environmental justice goals into political tokenism.[25]

A second dimension of the incorporation and transformation process can be apprehended by examining the movement's goal shift during the 1990s. Organizations that had originally focused on helping communities to fight toxic exposure increasingly recognized that the logic of NIMBY had resulted merely in displacement of waste and pollution to other communities. Often the displacement had an unfortunate politics of class and race, because white middle-class communities were better able to win NIMBY battles, and low-income ethnic minority communities were not. As Robert Bullard noted, NIMBY could lead to PIBBY ("put in blacks' back yards"), and likewise even the successful struggles of African-American communities could lead to a downward shift of toxic burden to Native American communities or the export of waste to low-income countries. As a result anti-toxics and environmental justice organizations increasingly articulated a goal of NIABY ("not in anybody's back yard").[26]

In addition to a shift toward NIABY, the goals of the environmental justice movement also diversified from the remediation of toxic exposure to broader issues associated with inequality and the general built environment. A position statement in the wake of the 1999 National People of Color Environmental Leadership Summit included seventeen principles that were already quite far reaching, and a second statement issued after the second summit in 2002 indicated a considerable broadening of the agenda. By 2002, the vision was more multi-ethnic and global in scope, with topics including climate justice, trans-boundary waste trade, economic globalization, radioactive colonialism, and biopiracy. Likewise, the vision was broadened from a focus on waste and toxic exposure to include the call for "equal protection" under health, employment, housing, transportation, and civil rights laws.[27]

In summary, there were two major IOMs oriented toward industrial waste and the pollution of industrial manufacturing: the post-preservationist, mainstream, national environmental organizations, with their focus on national environmental policy, and the environmental justice/anti-toxics movements, with their focus on local toxic exposure and the non-local, even global, issue of unequal exposure. On waste and pollution issues the mainstream environmental movement has in some cases adopted the goal of a moratorium, but its strategies have shifted in response to the rise of neoliberalism toward regulations based on market-based policy instruments such as emissions trading. Although it is important to have a watchdog that at the minimum can alert the public to a government that tends to sacrifice sustainability and justice goals on the altar of corporate profitability, the watchdog has often been likened to a lapdog. The incorporation and transformation process is highly developed in this wing of the environmental movement, but many of the organizations never had a history of grassroots, protest-based opposition. By forsaking the mass mobilizing potential of the politics of the moratorium, this wing of the environmental movement has restricted its work to incremental policy improvements and, increasingly, the rearguard action of attempting to stop retrenchment. Some analysts, including Michael Shellenberger and Ted Nordhaus, have laid the failure at the feet of the mainstream environmental movement and claimed that it lacks a broader political vision that is consistent with core American values such as entrepreneurship and technological innovation. Although their critique does point to the general weakness in the IOM strategy if it is not wedded closely to a TPM strategy, the TPMs associated with clean energy and clean production have also met with resistance.[28]

Notwithstanding the many and well-known weaknesses of the national environmental organizations oriented toward industrial pollution, they can claim that their continued vigilance has resulted in at least some changes in the technological field of industrial manufacturing. Changes include the reduction in industrial emissions of air and water pollution through emissions technologies and changes in production practices, sunsetting of some of the most dangerous chemicals, reduction of solid-waste leaching through new landfill technology, and restoration of some ecologically damaged brownfield sites. However, the examples of environmental remediation and amelioration have been accompanied by general increases in consumption and waste generation, growing evidence for the atmospheric and climatic effects of greenhouse gases, and new risks associated with ongoing innovation in the chemical industry, including nanotechnology. Furthermore, improvements in air and water quality at an aggregate level have coincided with the increased concentration of pollutants in some regions, including in neighborhoods with higher percentages of the poor and ethnic minority groups. Notwithstanding the legislative legacy of the 1970s and even the emissions trading innovations of the 1980s, in some parts of the country air quality since that time has declined, and chlorine-based chemicals continue to permeate products, the waste stream, and the biosphere. Although the technological field regarding waste and industrial pollution has changed dramatically since the 1960s, and the mainstream environmental organizations have played a role in enforcing regulations and preventing rollbacks, the general problems of environmental damage and toxic exposure from industrial production are far from being solved.[29]

In contrast to the mainstream environmental movement, the environmental justice movement has become the driving force of opposition to waste and pollution. Like the grassroots anti-CAFO mobilizations and the regional alliances of the anti-nuclear movement, the environmental justice movement draws on the powerful mobilizing potential of the geographical proximity of communities to environmental and health hazards, and at the same time it has successfully tapped into the legacy of the civil rights movement. J. Timmons Roberts and Melissa Toffolon-Weiss note that mobilizations achieved greater success when the opponent was a private company with a new siting, there was strong support from national organizations and extensive press coverage, and the environmental justice groups were represented by public interest lawyers. In other words, it may be possible to specify with some degree of precision the conditions under which local environmental justice struggles are successful and unsuccessful.[30]

As the environmental justice and anti-toxics movements have achieved local victories and shifted toward national and international arenas, the localist strategy of NIMBY has tended to shift toward a politics of NIABY. In making the transition, the movement will end up shifting into the TPM politics of the design of industrial technology and the greening of production processes, because the only way to avoid waste displacement from one community to another is to reduce waste and pollution in the first place. As I will argue in the next chapter, during the 1980s some recycling advocates and environmental justice activists converged over the issue of stopping incinerators. A second convergence of environmental justice goals with the TPM goals of zero waste production and the redesign of products and production processes is likely to emerge as the environmental justice movement widens its political analysis and its articulation of goals.

In support of the argument that environmental justice organizations will confront the politics of design, there is already some convergence between environmental justice organizations and sustainability efforts at a local level. As Julian Agyeman has emphasized in his discussion of "just sustainability," some local environmental justice organizations have undergone diversification and mission shift toward local sustainability projects. For example, he examined how Alternatives for Community and Environment of Roxbury, a low-income district of Boston with a mostly Latino and African-American ethnic composition, has worked not only to reduce and monitor air pollution, clean up a brownfield site, and stop an asphalt plant but also to green the public transit system. Another example, closer to where I live, is in Arbor Hill, an African-American neighborhood in Albany, where a successful environmental justice campaign against a waste incinerator resulted in a settlement that funded the Arbor Hill Environmental Justice Corporation. The organization not only monitors ongoing pollution and environmental health issues but also works on restoration projects for a local watershed area. The examples of the organizations in Boston and Albany give a sense of the trajectory that local environmental justice organizations can undergo as they diversify their missions from battles to end egregious toxic assaults on a community to more general greening and restoration projects.[31]

The Boston organization's campaign against diesel bus pollution is also an example of a broader set of campaigns that have occurred in other cities, where environmental justice concerns have become linked to the politics of alternative technologies. In Boston, Los Angeles, New York, and San Francisco environmental justice coalitions joined with public health groups to demand an end to old, dirty diesel buses and to the practice of locating diesel bus yards in low-income neighborhoods. The groups originally advocated compressed natural gas (CNG) as an alternative technology, and in some cities their mobilizations influenced decisions by transit agencies to purchase CNG buses.. However, the threat of diesel conversion to CNG and the tightening of federal and state emissions regulations pushed the diesel industry to change its emissions technologies. By the early 2000s the diesel industry had responded by developing a variety of design innovations (e.g., hybrid-electric engines, particulate traps, and ultra-low sulfur fuel) to reposition diesel as meeting the clean technology goals. In New York, Boston, and some other cities, transit agencies opted to shift back from CNG to "clean diesel." In other words, the environmental justice mobilizations played a role in the shift of local technological fields toward reduced emissions, but in some cases their original demands to "dump dirty diesel" were incorporated and transformed into fleet purchases for clean diesel.[32]

Regarding the general issue of the relationship between IOMs and a technological field, one can see the most dramatic effect at a local level in the aftermath of a successful environmental justice campaign. Where incinerators, chemical plants, diesel bus barns, and old diesel buses have been closed or taken out of service, the material culture and environment of a neighborhood are dramatically affected. Furthermore, opportunities emerge for brownfield conversion and the creation of new businesses that provide jobs through a production process that does not subject the community to high levels of health and environmental risk. At the national level, however, neither the environmental justice movement nor the mainstream environmental organizations can tell a story similar to that of the anti-nuclear-energy movement of the 1970s, when the growth of an entire industry was brought to a halt. Instead, to the extent that the movements can claim to have shifted the technological fields of industrial manufacturing and waste, the changes have been incremental and located at the edges of production processes, production materials, and emissions technologies.

Infrastructure

Having witnessed the high speed and efficiencies of the German autobahn during World War II, American military and political leaders supported the development of the interstate highway system as a central infrastructural investment for the postwar era. Any number of design modifications could have been introduced to result in a much more sustainable and just scale expansion of the transportation infrastructure. Highways could have stopped at the outer limits of urban areas, where bus, train, and light rail connections could have integrated intra-city and inter-city public transportation. Highway rest stops could have been constructed as transportation hubs that linked passengers to buses, trains, and urban transit systems. Instead, the automotive industry supported the destruction of streetcars and their replacement with buses, while the interstate highway system bulldozed urban neighborhoods, created riverine divisions in the urban landscape, displaced both inter-city and intra-city rail, consigned much of inter-city and intra-city bus use to the poor, and offered the automobile as the only feasible means of connection between the suburbs and the rest of the world.[33]

Almost as soon as the highway system had taken off as a public works megaproject, there were significant mobilizations against highway construction plans. In the same dynamic of NIMBY to PIBBY discussed in the previous section, the success of middle-class neighborhoods at stopping highways sometimes led to diversions of highway projects into low-income neighborhoods. As with environmental waste sites such as incinerators, bus yards, and chemical manufacturing plants, the low-income neighborhoods were less able to mobilize the resources needed to stop the construction. In some cases, such as the Overtown neighborhood of Miami, highway construction destroyed a vibrant, economically healthy African-American neighborhood. In other cases, such as San Francisco and Boston, a diverse range of neighborhood and ethnic coalitions mustered enough power to stop highway construction through significant parts of the city. Anti-highway mobilizations also emerged in other cities during this period, including Baltimore, Milwaukee, and Seattle.[34]

Anti-highway mobilizations tended to be very local in orientation, and they never achieved the national scope and duration associated with the anti-nuclear-energy movement. The anti-highway campaign in Boston provides some insight into the structure and dynamics of this type of industrial-opposition action. The conflict involved a long-standing battle between grassroots groups and government officials over plans to build an inner belt and to extend various interstates. In 1966 a group of reformist planners and activists, including a salaried employee of Students for a Democratic Society, formed Urban Planning Aid. Two years later the group merged with other community groups to form the Greater Boston Committee on the Transportation Crisis, which became the main coordinating organization for groups from around the region that were opposed to highway construction plans in their neighborhoods. Similar to the trajectory in the anti-nuclear-energy movement, the general political goal evolved from supporting alternative routes and construction designs to a call for a complete moratorium on highway construction within Route 128. The coalition also developed to include a wide range of neighborhood groups, progressive elected political officials, and experts and academics from local universities. Eventually, the Boston City Council and mayor embraced the call for the end of construction, and in 1970 the Republican Governor and former state transportation department official, Francis Sargent, also accepted the moratorium, albeit reformulated as a partial moratorium with his proviso that one of the highways, Interstate 93, would go ahead. As also occurred in San Francisco, the Boston mobilization was successful because it had achieved a broad, multi-ethnic coalition that effectively worked opportunities at the local, state, and national level.[35]

The anti-highway mobilizations of the 1960s and the 1970s were faced with the classic weakness of oppositional politics: they needed to provide more than opposition in order to maintain credibility. The most obvious solution to the problem of traffic congestion other than highways was public transit, and the Boston coalition's mobilization represented one of the beginnings of the shift in the allocation of federal highway funds to public transportation projects. During the 1960s there was some federal funding for public transportation under the Urban Mass Transportation Act, but the amount was small. The anti-highway organizations focused increasingly on shifting federal transportation funding to include more public transportation. The goal shift is a point of connection both with the urban planning reform movements to be discussed in the next chapter and the more general anti-sprawl movement that is the legacy of the anti-highway movements of the 1960s and the 1970s.[36]

In the 1990s and after, the opposition campaigns directed at urban infrastructure underwent two major changes. With many of the urban highways already built or cancelled, action shifted as middle-class suburban homeowners turned their attention to other forms of growth and development. One area of mobilization against new infrastructure is a product of one of the most visible faces of globalization: the consolidation of the retail industry and the elimination of wholesalers through the creation of the warehouse-style or "big box" superstore. After a successful campaign in 1993 to stop the location of Wal-Mart in Greenfield, Massachusetts, the organization Sprawl-Busters was founded to help communities stop the construction of "big box" stores. Within 10 years the organization could claim about 250 campaigns around the country. There may be even more campaigns against the expansion of airports and the attendant problems of congestion and noise pollution, and since 2000 there have been more campaigns against cell-phone towers. However, at this point the social-science and historical literatures are undone, and we will have to await future research to understand better the dynamics of such campaigns and their patterns of incorporation and transformation.[37]

The oppositional infrastructure campaigns today have a local orientation that is continuous with the anti-highway campaigns of the 1960s and the 1970s. The class politics have shifted from multi-class urban neighborhood coalitions to coalitions that unite independent retailers with middle-class suburban homeowners who are concerned with the effects of additional growth on property values, congestion of roads and schools, and quality of life in general. The shift in class address is of interest because in some ways it is in the opposite direction from the shift that occurred with industrial opposition from the more middle-class, mainstream environmental movement to the environmental justice movement. My hypothesis for the difference is that the rate of highway construction in urban areas has slowed dramatically, whereas the construction of "big box" superstores and sprawl has continued unabated. Broad coalitions with middle-class participation, such as were found in some of the successful anti-highway campaigns, appear to be more successful than concentrated opposition from single neighborhoods.

Again, effects on infrastructure as a technological field can be dramatic but are generally highly localized. The transformation of opposition to highways into support for public transit has had some more general effect on the development of federal funding in support of urban rapid transit, but transit use in the United States remains much lower than in most other wealthy countries. Although growth in highways has slowed, the growth in other forms of infrastructure continues unabated in the outer metropolitan areas. Even where local mobilizations are successful, the new infrastructures are able relocate with ease to another community in the metropolitan area.

Finance

In the financial field there has been considerable consolidation of local banks but relatively little in the way of public mobilization against the process. Rather, the primary IOM in the financial field—the anti-corporate, anti-globalization movement—began in the global South in opposition to the policies of global financial institutions, such as the structural adjustment policies of the International Monetary Fund. In many countries the movement called for a moratorium on the national debt, and national governments achieved a partial moratorium through debt rescheduling. Environmentally oriented opposition also arose with respect to the damaging showcase development projects funded by the World Bank. Projects such as large hydroelectric dams that displaced indigenous and rural villagers often had disastrous social and environmental consequences, suffered from corruption that severely weakened credibility for the high-tech development paradigm, and left the governments of low-income countries saddled with debt and high interest payments. In addition to the effects of projects supported by the World Bank, the expansion of trade liberalization caused increasing dislocations among the less fortunate, including small farmers.[38]

In the 1990s farmers and workers in the global South led protests over trade liberalization, and by the end of that decade the protests against the World Trade Organization had spread to the North. The Northern protests tended to be more diverse, and participation from students, environmentalists, and the activists was greater. The separate strands were interwoven in a shared analysis of the subversion of democracy wrought by global financial institutions and corporate control over governments. Although there were points of convergence among Northern and Southern participants in the anti-corporate, anti-globalization movement, significant differences of emphasis remained both between and within the two hemispheric wings of the movement.[39]

In a vast field of action that includes finance, the field of fields, and the anti-corporate, anti-globalization movement, the movement of movements, this section will focus on one aspect of oppositional politics in the United States: the environmental dimension of anti-corporate activism and its relations with the other efforts at social change regarding investment and its governance. Within that narrow topic, the discussion will again attempt to characterize the incorporation and transformation of oppositional action and the shifts in the technological field.

Regarding the global institutions of finance, in the 1980s environmentalists became concerned with the connection between World Bank policies and environmental destruction in developing countries. Of particular concern was the construction of roads and dams that led to rapid colonization and deforestation of previously protected or at least relatively undisturbed areas such as the Amazonian rainforest. In 1983 the National Wildlife Federation, Environmental Policy Institute, and Natural Resources Defense Fund led a coalition that opposed the environmental policies of the World Bank and other multilateral lending agencies. In response to the campaign, but also as a result of mobilization in the developing countries, in the late 1980s the World Bank began to incorporate environmental considerations into its funding programs. By 2000 the Bank claimed to be the leading lender for environmental projects in the developing world, and it had also shifted its mission to include social justice considerations. Critics remained unconvinced that the Bank's changes were as profound as its reports and goal statements suggested; in turn, at least one Bank official stated that activists had overestimated the influence of the Bank and were, in effect, barking up the wrong tree. As in other cases, the incorporation of IOM goals was incomplete, and the politics of a moratorium (especially on targeted development projects) were transformed into the politics of environmentally oriented development.[40]

Whereas the shift in policies at the World Bank is sometimes held up as an example of successful mobilization of environmentalists directed at global financial institutions, the same cannot be said for the World Trade Organization. The World Bank increasingly reached out to NGOs, and the Bank also selected organizations in the NGO field by favoring "operational" rather than "advocacy" organizations (a pattern of transformation toward service provisioning that, I will argue latter, is widespread in the access fields). In contrast, the World Trade Organization (WTO) failed to develop deliberative arrangements with the NGO sector. Some reformist environmental NGOs developed contacts with the WTO Secretariat through representatives in Geneva, but leading environmentalist groups (including Greenpeace, Sierra Club, and the Rainforest Action Network) were so roundly rebuffed that they participated in the 1999 Seattle protests and other anti-WTO protests. In the wake of various WTO and North American Free Trade Agreement rulings that challenged or overturned national environmental legislation, environmental organizations have attempted to build coalitions with labor and other groups to reform the transnational trade organizations.[41]

In addition to mobilization in favor of changes to global financial institutions, activists have also developed campaigns against the environmental policies and investment practices of transnational corporations. Direct anti-corporate mobilization can be divided into two types: campaigns against pollution and toxic waste as in the environmental justice movement (or against waste in zero-waste campaigns), and campaigns against the extractive practices of private companies. This section will focus on the latter, with the understanding that continuities with the environmental justice movement should not be forgotten. At the same time, as the following cases will demonstrate, there are also continuities with the older, preservationist wing of the environmental movement. In a study of anti-corporate boycotts, the two the largest groups were the Earth Island boycotts on tuna and the Rainforest Action Network boycotts on timber; they are discussed here.[42]

One of the best-known cases of environmentalist mobilization in the United States directly aimed at corporations involves tuna harvesting techniques that result in unnecessary dolphin deaths. Because dolphin herds often swim above tuna, the rate of dolphin deaths associated with tuna harvesting increased as the technologies of net-based tuna fishing improved. A campaign led by Earth Island Institute in the late 1980s resulted in a boycott of the H. J. Heinz Company, which owned market-leading Star-Kist Seafood. In 1990 the three major tuna-processing companies responded to the campaign by pledging to sell only dolphin-safe tuna, and the federal government passed legislation to support a dolphin-safe label for tuna sold in the United States. The legislation led to an embargo of tuna from countries that did not use the dolphin-safe technology, and the aggrieved countries filed a complaint with the General Agreement on Trade and Tariffs (GATT), the precursor organization to the WTO. In 1991 and 1994 GATT ruled in favor of Mexico and the European Union on the embargo issue. A subsequent compromise measure that was worked out with twelve tuna fishing countries ended up dividing environmental organizations, and the issue remained unresolved 10 years later. However, in direct parallel with the mobilizations against genetically modified food, the dolphin-safe-tuna activists were more successful with respect to retail food corporations. Those companies were willing to make changes to avoid consumer boycotts, brand dilution, and loss of market share. In contrast, the federal government found itself reversing laws based on the rulings of GATT and the WTO.[43]

Another example of action against corporate environmental policies and investments is the Rainforest Action Network's campaigns against corporations that harvest tropical hardwoods and lumber from old-growth forests. In the early and mid 1990s the organization carried out a boycott against Mitsubishi's tropical hardwood harvesting operations, and in 1998 the corporation agreed to shift to sustainable forestry. Since that time RAN's campaigns moved on to home supply companies and logging companies in the United States, and beginning in 1999 RAN and other environmental organizations began to win victories from supply companies to stop the sale of old-growth lumber in their stores. The campaigns also expanded to include companies that use old-growth lumber in paper products.[44]

In its subsequent Global Finance Campaign, RAN targeted bank investment policies on environmental issues. After Citigroup (the industry leader) rebuffed RAN's efforts to undertake negotiations, RAN inaugurated a credit card boycott in 2000 and began organizing events, often on college campuses and with the support of college students. After three years of campaigning Citigroup announced a new industry standard for environmental investing, and shortly thereafter Bank of America responded with a similar policy. The new policies restricted investment for extractive projects in ecologically sensitive areas, promoted sustainable forestry and renewable energy, and assessed investments related to greenhouse-gas emissions. RAN continued to develop new goals and to expand the campaign to other banks.[45]

Regarding the issue of how the IOM changes, the RAN campaigns suggest a three-phase pattern: an initial phase of attempted negotiation, when the organization is dismissed; a second phase of confrontation, when the organization utilizes protest tactics and boycotts; and a third phase of partnership, when the organization makes a move to incorporate the environmental goals. In the third phase at least one corporation may break ranks and implement a version of the activist organization's call for a moratorium on a specific type of investment. Sometimes the NGO will enter into certification schemes for alternatives, and sometimes it will legitimate the use of third-party certification. However, the industry may also respond with its own version of a certified product. For example, in 1993 the World Wildlife Federation and Greenpeace launched the Forestry Stewardship Council, but the timber industry responded with weaker standards in the Sustainable Forestry Initiative. Only after ongoing pressure from environmental organizations did the industry develop stronger standards and third-party certification.[46]

In the three cases discussed here (the World Bank's development policies, tuna harvesting, and anti-corporate Rainforest Action Network action), the target organizations incorporated the call for a moratorium on specific types of development projects, fishing technologies, and unsustainable investment patterns. In the place of old technologies and products the targeted organizations developed new designs for their technological systems—environmentally oriented development projects, dolphin-free tuna labeling, and certified lumber products—that shifted the politics of industrial opposition into one of industrial innovation. In terms of my movement categories, there has been a shift from the ideal typical world of the IOM to that of the TPM. The politics do not disappear, but they become transformed from a focus on the degree and scope of a moratorium to object conflicts that center on the design of the innovated technology or product.[47]

Regarding this brief survey of mobilizations against global financial institutions and global corporations on environmental issues, a few tentative patterns in the way of hypotheses can be generated. As in the anti-pesticide and anti-highway mobilizations, the call for a moratorium on a particular type of technology and productive practice—such as extractive projects in oceans and rainforests—has been relatively successful, especially where firms are concerned with the effect of media coverage on retail brands, such as tuna-processing companies, building supply companies, or credit cards. Again, the targeted moratorium does not require a fundamental change in ownership patterns or even market position; rather, it requires a shift of both oppositional organizations and corporations from the contentious politics of protest to the partnership world of the TPM. The environmental organization may undergo a shift in mission from opposition to certification, at least for the relevant campaign. As the change occurs, there are various shifts in technological fields—the design of infrastructure development projects for the World Bank, and fishing and forestry technologies for corporations and private banks—that suggest how social movements and industries can become intertwined in processes of industrial innovation.

Conclusions

On first analysis the dramatic but limited and narrow successes of the David-and-Goliath politics of the IOM pathways may be surprising. One source of strength is the relationship with TPMs. As I have argued, the moratorium can be delegitimized if is not connected with an articulation of an alternative, just as the motivation for embracing the alternative is powered partly by the perceived flaws in existing industrial practices. In this sense the IOMs and TPMs presuppose one another and are ultimately separable only as analytical categories. There is also a more direct relationship that emerges when the IOMs target corporations, seek changes in corporate investment decisions, and manage to shift industrial resistance to negotiation. When the breaking point is reached, the industrial-opposition strategy also shifts to the politics of innovation through the object conflicts that emerge over various certification schemes and standards. Examples discussed above are emergence of sustainable lumber standards and acceptable tuna fishing standards, but they also include negotiations over which type of targeted technology and product is acceptable and which type is not. Opposition shifts from a complete moratorium to a more complicated negotiation over design. Which type of targeted product and technology—genetically modified food, pesticide, factory farm design, nuclear energy waste site, pollution remediation technology, fishing technology, and forestry practice—will be counted as acceptable and which type will not?

Another type of connection across pathways occurs with the localization of opposition politics. A threat to a geographically restricted community represents a tremendous opportunity for mobilization. Leaders of small opposition groups can draw on existing local networks and organizations to form broad, local coalitions that in turn can find support from national organizations looking for on-the-ground connections with present and potential members. In addition to the wide range of environmental justice cases, examples include local residents who mobilize to maintain quality of life threatened by infrastructural development such as factory farms, nuclear power plants, highways, or cell phone towers. Another example is mobilizations by farmers who are threatened by cross-pollination from GM food. However, a purely localist strategy, without a concomitant strategy focused on higher-level governmental institutions and transnational corporations, is likely to result in local victories that merely displace hazards and risk to other, less well-organized (and often poorer) communities.[48]

A third type of connection and opportunity for coalitions emerges when the call for a moratorium on a particular industry can be framed as a rights issue, and as a result the IOM campaigns can develop a point of continuity with the human rights politics of access to goods and avoidance of bads. Examples include the rights of allergic individuals to food free of new allergens or of children to have cleaner air on school buses. Those campaigns draw on widely held values of justice and the rights of people not to be unfairly exposed to toxic risks. One might even extend the argument to the rights of animals, such as dolphins in the anti-tuna-fishing campaigns.

The potential for coalition building is only one factor that could explain the seemingly improbable successes that sometimes occur in the IOM campaigns. During the 1970s federal and state governments were more amenable to adopting legislative approaches to environmental remediation, whereas during and after the 1980s political opportunities closed down, unless a significant public health risk could be documented. But even when political opportunities shrank, the IOMs could still take advantage of divisions within government, such as across levels of government and across agencies and branches at a given level. They could also take advantage of divisions within industries and recruit whistleblowers who sided with movements on crucial issues (as in the anti-nuclear-energy movement) as well as firms or industrial segments that were more willing to shift in response to movement demands. Where activists were able to generate significant risk of brand dilution, they were able to leverage a change in the targeted industry. The strategy probably works less well where consumer choices cannot be brought to bear on a retail industry, such as electricity production through nuclear energy before the era of retail choice.

Although the discussion of coalition building and political opportunities is helpful as a way of understanding the conditions under which some IOM campaigns are more successful than others, my primary problem has been to understand the general patterns in the history of the industrial opposition campaigns and movements since the middle of the twentieth century. My argument is that the IOMs achieve some successes, but the pattern of success is through a process of incorporation and transformation. Political and/or economic elites respond, at least in several of the cases, to movement demands by accepting some of the demands (by incorporating those demands into new policies, production practices, and technology design) but also by transforming them. For the IOMs the incorporation and transformation pattern generally involves the politics of the partial moratorium. For example, a call to ban all GM food ended as a ban only on foods that are most likely to become deadly allergens for a small percentage of the population, and the call to end all nuclear power ended as a moratorium on new construction and an agreement to eliminate some of the most unsafe designs. Likewise, in response to anti-highway mobilization in Boston, the governor embraced the grassroots demands but did not cancel all the highways; the moratorium on new highway construction was only partial. To some degree, the use of a partial moratorium merely refashions the old strategy of divide and conquer in the new arenas of the politics of technological and infrastructural development and the investment decisions of political and industrial elites. They tend to select from the goals of the oppositional groups and make enough change to take the steam out of the movement.

That there is nothing new here should not be cause for disappointment; rather, one should remember that the contours of the technological field are altered by the struggles. The design of genetically modified food, the use of pesticides, the availability of nuclear energy, the location of emissions control technologies, the levels of industrial waste and pollution, the siting of highways and superstores, the availability of urban transit, and financing for environmentally destructive projects would all be different today if there had not been mobilizations by the IOMs to oppose the existence of, or specific forms of, such aspects of technological fields. Processes of incorporation and transformation take place, and the IOM may lose steam in the wake of such processes, but the next wave of mobilization takes place in a historical terrain that is changed both politically and technologically.

5

Technology- and Product-Oriented Movements

The politics of the moratorium found in the industrial opposition movements creates a legitimacy gap, because the opposing groups are vulnerable to claims that their politics are merely negative and to claims that the groups have failed to articulate an alternative. Consequently, the oppositional type of action discussed in the previous chapter works best when integrated with action that articulates and develops the alternatives. However, the synergetic effects go both ways. Technology- and product-oriented movements (TPMs) rely on IOMs to motivate funders, to help establish a hospitable regulatory environment for new firms and industries, to supply young recruits who seek careers and jobs in emergent industries, and to educate consumers about the value of buying alternative products.

Many organizations integrate oppositional and alternative politics, and there is often considerable traffic between IOMs and TPMs, which in the end are only theoretical concepts that can help guide the analysis of historical patterns. Likewise, the democratic goals of local control and access for the poor, which are discussed in the next chapter, may also be included in organization missions. Nevertheless, IOMs and TPMs can be distinguished as ideal types. Unlike IOMs, TPMs are oriented toward developing alternative systems of technology and products, as well as making related changes in markets, industries, and regulations. Because the fundamental goal is to develop alternative industrial technologies and products, TPMs are compatible with societal transformation in ways that are quite consistent with elite prescriptions for societal change, such as through technological innovation, the development of new markets and industries, and state subsidies for new industrial development. The compatibility with business as usual is correlated with the social address of TPMs, which tend to draw their strength from the middle class and especially from reform movements within professions and industries. In this sense, TPMs can be compatible with the argument, associated with ecological modernization, that the various environmental crises can be resolved without making fundamental changes in economic institutions. The tendency of TPMs toward moderate politics is likely to be uncomfortable not only for some activists in the IOMs and localist groups but also for some scholars of social movements, and it may account for the relatively understudied and under-conceptualized place of TPMs in the literature on social movements.

In addition to a difference in goals, the repertoires of action and organizational form of the ideal typical IOM and TPM differ. TPMs sometimes engage in protest politics; however, as has sometimes been noted for new social movements, they also channel social-change action through the construction of alternative institutions. Material culture and the institutions that create and diffuse it, rather than street-based protest and other confrontational repertoires of action, are typically the primary means of social change. Organizational structure also differs significantly from IOMs. Although IOMs may work with professional reformers, such as the scientists who documented risks and hazards in the nuclear energy and weapons industries, for the TPMs the reform movements are crucial and often even the driving force. In this sense the IOM and TPM are inversions of one another.

Within the TPM there is also a division between a social-movement side characterized by NGOs and a private-sector side of reformist firms. The social-movement side of TPMs may see private-sector partnerships as a vehicle for their politics, and the firms may also draw on the movement side for support. Furthermore, one can also find the hybrid figure of the activist-entrepreneur, especially during the early phase of the TPM's development. Consequently, just as there is a symbiosis between the IOMs and the TPMs, so within TPMs there is a private-sector symbiosis, or a mixture of civil-society organizations with an advocacy orientation and for-profit firms that develop and diffuse the alternative technologies and products. Because of the symbiotic relationship between advocacy organizations and the private sector, the ideal typical TPM has a dual organizational structure: it has a social-movement wing, which organizationally consists of NGOs and related organizations that are often connected to IOM politics, and a reform movement wing, which consists of firms, professional reform organizations, and associated scientists and designers that pioneer, develop, and market reform-oriented technologies and products. Often scientists and inventors face their own battles of marginalization.

For TPMs the incorporation and transformation process by governmental and industrial elites operates less around the politics of the partial moratorium than the politics of complementarization and object conflicts. In some cases the alternative technologies and products are first created by small-scale entrepreneurs who operate in a movement-like atmosphere. They may build linkages to a countervailing industry that plays a crucial role in taking up some of the alternative products. An example is some energy companies, which supported renewable energy at a time when the utility companies were still worried about the threat that alternative energy posed to grid-supplied energy. As the TPMs achieve success, the targeted industries often begin to show an interest in incorporating or co-opting the alternatives, and in the process the design of the technologies and products undergoes a transformation. The usual direction of the transformation of the design involves "complementarization," that is, modification of aspects of the design that are in conflict with the dominant technologies in an industrial field, so that the alternative becomes complementary. An example is the transformation of community-controlled or off-grid renewable energy into grid-controlled wind farms run by large utilities. The process of incorporation and transformation may coincide with mergers and acquisitions in the private sector, particularly if the for-profit firms that originally supported the alternative technology and product were small-scale and entrepreneurial. Their need for capital, or simply the need of aging entrepreneurs to cash out and retire, also drives the incorporation process at the organizational level.[1]

The process of incorporation and transformation tends to heighten object conflicts, that is, definitional struggles over the technology of production and final product. Ultimately the struggles are about the general contours of the technological field, such as the mix of renewable versus fossil-fuel energy, or grid-supplied versus distributed energy. The object conflicts that occur during the incorporation and transformation of TPMs tend to take place between, on the one side, the social-movement side of the TPM and the small-scale firms that developed the original alternatives, and, on the other side, the large industries, together with the small firms that have been acquired. As the incorporation and transformation process develops, the primary political issue shifts from the conflict between alternative and existing technologies/products to one between various types of alternatives and the various degrees of cooptation that are occurring as the alternatives become mainstreamed.[2]

Food and Agriculture

As the IOMs developed against pesticides, genetically modified food, and factory farming, they pointed to organic food and agriculture as an alternative to the health and environmental risks associated with industrial agriculture. For example, the Pesticide Action Network shifted from oppositional politics to support for sustainable agriculture, and an influential California organization, the Community Alliance with Family Farmers, broadened its goals from support of farm workers and opposition to pesticides to support of sustainable agriculture. The mission statements and program descriptions of other environmental organizations involved in oppositional food politics also reveal a similar mix of oppositional and pro-alternative-agriculture goals. Organic agriculture provided an alternative that could allow the opposition campaigns to move beyond the politics of the moratorium.[3]

However, the origins of the organic food and agriculture TPM predate the late-twentieth-century campaigns against pesticides and genetically modified food. The development of the organic food and agriculture movement emerged hand in hand with the urbanization of American society and the industrialization of agriculture, and its earliest manifestations can be traced back to the 1920s and the 1930s. Most historical accounts view late-twentieth-century organic agriculture as the direct legacy of various early- and mid-twentieth-century European thinkers. One convenient point of origin is the mid 1920s, when Rudolf Steiner, the leader of the spiritually oriented anthroposophy movement, taught a course on biodynamic agriculture. At roughly the same time the British colonial scientist Sir Albert Howard began an agricultural research station in Indore, India, where he improved a composting method and an approach to agriculture that became known as the Indore Process. The

1940 British book *Look to the Land* is said to contain the first use of the term "organic" in the sense similar to its usage today. By 2000, an organic movement had given birth to an organic industry, and social-science and historical literatures had developed on the politics of organic agriculture.[4]

In the United States during the middle of the twentieth century the leading advocate of organic farming and gardening was J. I. Rodale. In 1941 he came across Howard's work, and shortly afterward he bought a farm and launched the magazine that later became *Organic Gardening and Farming*. He sent thousands of copies of subscription solicitations to farmers, but, in an indication of the lack of environmental consciousness at that time, he received few responses. He eventually found a supportive niche among gardeners, who were often European immigrants of his generation. Rodale also attempted to interest scientists in testing and developing organic agricultural research, but the research community ignored his requests and even responded with attacks on the idea and technology. His response to their attacks drew attention to the role of chemical companies on the boards of agricultural colleges, predated environmentalist critiques of interested science, and provided an early example of the devolution frame, that is, the argument that scientific research had foresworn its public benefit and sold out to industrial interests. Given the lack of interest in organic agriculture research from most major universities, the knowledge of organic food and agriculture during its formative years was, like the experimental science of the seventeenth century, based in institutions outside the university setting. Radical in terms of its challenge to contemporary agricultural knowledge, Rodale's original vision of the object "organic" was also quite technical and relatively free of the concerns with justice, sustainability, and localism that would later preoccupy the activist end of the movement. In an indication of the health-oriented individualist politics of his concept of organic farming, the mainstay publication of the Rodale business during the early decades was *Prevention*, and at one point Rodale even considered folding *Organic Gardening and Farming* into the larger magazine.[5]

It is worth dwelling for a moment on Rodale's publications and publishing company, because it provides a good example of the hybrid nature of TPMs, where social movements and business development meet and mingle. With the blossoming of the counterculture, in 1969 and 1970 a younger generation began to flock to organic food and farming. Subscriptions to Rodale's magazine skyrocketed, and by 1971 a long article in the *New York Times Sunday Magazine* referred to organic as a movement. In 1969, two years before his death, Rodale responded to growing environmental concerns by launching the *Environmental Action Bulletin*. Reflecting emergent concern with recycling, the company also added the subtitle "Journal of Waste Recycling" to its publication *Compost Science*. Rodale's son, Robert Rodale, developed the environmental and social side of the emerging organic movement, and under the son's guidance the relatively technical and health-oriented organic vision of the senior Rodale merged into the 1960s social movements, including the "back to the land" movement. The family's publishing company made money as a business even as it also served a political role as an advocate for a more environmentally oriented version of agriculture.[6]

The Rodale organization helped to define the developing organic movement/industry in another way: it offered the first certification program and promoted the idea of organic farming standards. In turn, a new wave of organic farming organizations developed organic standards. The Northeast Organic Farming Association, which was founded in 1971, may have been the first organization of organic farmers in the country. However, similar organizations appeared in other states during the next few years, and California Certified Organic Farmers led the development of organic standards. Although the original California group has been described as consisting of hippie farmers who were interested in sharing information, it also pursued the goal of protecting the food quality claims of the small, organic farmers in the marketplace. Consequently, the organization pushed for state-level legislation, which was first passed in 1979. The move to focus the definition of organic on technical production standards, which to some degree was Rodale's legacy, rather than the more political issue of local ownership and consumption, opened the door for a divergence between the rapidly developing organic food and agriculture industry and the vision of sustainable local agriculture held by many of the countercultural, organic farmers.[7]

As the state-level organizations of the organic movement created the conditions for industrialization through the development of technical standards, they also shifted in a second direction beyond organic certification, technical assistance, and market development. Again, the changes are suggestive of the hybrid quality of TPMs as both industrial innovation networks and social movements. For example, the Northeast Organic Farming Association of Vermont developed a farm share program for low-income residents and seniors; the California organization developed environmental projects in support of salmon restoration, water conservation, and renewable energy; and the Northeast Organic Farming Association of New York had a medicinal herbs project, which represented another expression of the long synergy between organic agriculture and complementary and alternative medicine that Rodale had emphasized. The organizations also developed policy statements against genetically modified food and contributed to some degree to the food-oriented IOMs. Furthermore, the organizations associated with the appropriate technology movement (including the National Center for Appropriate Technology, the Center for Rural Affairs, and the Land Institute) also helped to build networks of sustainable agriculture advocates. In many of the organizations, social-change goals mixed with those of supporting a new industry, and projects were connected with other social-change action within the industrial field, such as anti-hunger work or opposition to pesticides and genetically modified food, as well as across fields, such as renewable energy and alternative medicine.[8]

The industrialization of organic agriculture occurred through two, parallel developments. In some cases, such as Earthbound Farms and Cascadian Farms, small alternative farms grew into large agricultural businesses. However, as research by Julie Guthman has now revealed, in the case of California (which, unlike many other states, was historically composed of large farms rather than small, family farms) much of the growth of organic farming occurred when large growers saw new market opportunities and made the business decision to convert part or all of their fields to organic production. She notes that the takeoff of the organic industry during the 1980s has been attributed to spikes of consumer demand driven by food safety scares that occurred with the pesticides Aldicarb and Alar. Many of the new entrants were mixed growers, and ownership patterns, at least in the case of California, were similar to those of conventional farms. Farms that started as organic tended to be smaller and to use organic production techniques more heavily (rather than producing to the minimum standard for certification), and they tended also to pay higher wages and have more local and direct marketing.[9]

As the organic food and agriculture industry developed, the industry diversified into processed organic food and non-food products such as clothing. Certified California Organic Foods became a battleground for object conflicts between the hippie "purists" and the more pragmatic mixed and large growers, and the latter sometimes selected other certification organizations that were more responsive to their vision of standards. In 1984 the larger growers joined with certification organizations to found what later became the Organic Trade Association. In 1990 the organization achieved its goal of establishing legislation in support of a national organic standard, and, as the industry grew, the organization also supported the diversification of organic agriculture from food into a wide range of organic products, including cotton and other textiles. Organic clothing and textiles lagged food but were growing at a rate of 20 percent per year during the early 2000s.[10]

With this background in mind, it is now possible to turn to my central interest, the incorporation and transformation process. In the case of organic farming the process can be broken down into three topics: farm consolidation, acquisitions of food-processing companies, and changes in the retail industry. Regarding farm consolidation, by 2001 several organic farms in California had expanded to the size of 2,000–5,000 acres, and the firm Horizon (purchased by Dean Foods) controlled about 70 percent of the American organic milk market. As the industry grew and consolidated, agricultural technologies underwent changes in the sense that the larger farms tended to be less committed to the full range of ecological farming techniques that were behind original conceptualizations of organic farming. The development of organic standards created a minimum level of production quality that was often below that of the smaller, locally oriented farms, and the standards ignored product quality.[11]

A second dimension of the industrialization process was the incorporation of organic food into the mainstream food-processing industry. In some cases, organic farms moved upstream into food processing, where profits were higher. In the United States the best-known case is probably Cascadian Farms, which was founded in 1971 to grow food for hippies in the region near the farm. Eventually the farm developed into a food-processing company and was acquired by General Mills. Many other organic food-processing companies have been purchased by the large food conglomerates, which decided to enter the profitable niche market through acquisition as well as through new product development.[12]

The case of Stonyfield is exemplary of the dynamics of TPM, so it will be examined in a little more detail. Stonyfield, a yogurt and dairy food company, was a member of a progressive network of small businesses called the Social Ventures Network, and it helped many small farms in the Northeast stay in business by converting to organic dairy production. Unlike Ben Cohen of Ben and Jerry's, who was also a member of the Social Ventures Network and who expressed some regrets about events that led to his firm's sale to Unilever, the owner of Stonyfield did not view the sale to the food conglomerate Dannon as selling out. When the leading goal is a technical and ecological one of conversion of the greatest amount possible of farmland to organic production, sale to a large conglomerate can be defended as an opportunity to convert more farmland to organic production and even to help the conglomerate undergo further greening. However, the sale may not be defended as easily on other grounds, such as protecting high wages and participatory structures for employees and price premia for organic farmers. In this sense the Stonyfield case exemplifies one of the central dilemmas of the entrepreneurial TPM firm. Although small, entrepreneurial TPM firms such as Stonyfield may embrace social reform values of employee participation and generous compensation packages, those values can be severed from the technical reform goals (e.g., increasing market share for organic products), particularly after incorporation in the form of acquisition by a large, publicly traded corporation. This is the Achilles' Heel of TPMs.[13]

In the retail industry the incorporation and transformation process has occurred less through acquisitions than through displacement by large corporations that entered the natural foods market. During the late twentieth century conventional grocery stores and chains (known as the "food, drug, and mass channel") were themselves undergoing significant consolidation, and by 2004 eight firms had 50 percent of the market, with Wal-Mart by far the largest supermarket retailer. By 2004 conventional supermarkets had also developed natural foods sections and organic foods offerings in about 73 percent of all stores. As of 2002, the conventional supermarket side of the natural foods retail market was growing more rapidly than natural foods retail stores (15 percent versus 9 percent), and its sales volume of $4.2 billion was already at a level of about half the volume of the natural food stores. Wal-Mart's 2006 decision to increase significantly its retailing of organic products will accelerate the trend. A second major change occurred within the category of natural foods stores due to the growth of natural food store chains such as Whole Foods and Wild Oats. The mainstreaming of organic foods retailing, through either the supermarket diversification of product lines or the development of natural foods retail chains, has coincided with the diversification of the organic foods category to include an increasing number of processed foods. The transition into processed foods tends to shift profits from organic farmers to food-processing firms and to increase price competition among organic food producers.[14]

On economic metrics organic food products remain only a small niche in the larger food industry, but even so the model of industrialized organic agriculture production with food sold through supermarkets has come to dominate the model of small farmers who market directly to consumers. However, several social scientists who have charted the transformation have cautioned against seeing the historical change as a simple replacement of an original social movement by the organic industry. Rather, they suggest that the growth of the organic industry coincided with parallel growth in farms and other institutions that blended the goals of sustainable agriculture, local control, and food security. By the 1990s one tended to find the terms "organic agriculture movement" or "organic food movement" replaced by terms such as "sustainable agriculture" and "local agricultural movement"; in other words, values associated with localism and sustainability had become increasingly salient. Had those values originally been encoded in the standards that defined organic, the history might have developed differently, although in the event of a conflict between codes and capital, one would expect that capital would find a way to change the codes.[15]

The differing views of the TPM for food and agriculture that have been developed in the organic food industry and the emergent movements in support of sustainable, local agriculture are the basis for a triangle of related object conflicts over the shape of the technological field. As was described in chapter 3, the definitional struggles over the object can be tracked in three areas: funding for research programs that shape the future of various types of alternative food, consumption decisions among the array of possible alternative food categories, and standards set by private-sector or governmental bodies that govern definitions.

Regarding research programs, one category of object conflicts involves funding decisions over how much money will be devoted to organic research. In my discussion of alternative pathways in science in chapter 2, I noted the low levels of funding for organic agricultural research in general. However, there are also issues regarding the design of research projects within the category of organic agricultural research. For example, because the category of "organic" includes a wide range of production technologies that may or may not include crop rotation, composting, and biological pest management, research agendas that focus on one dimension of the production technology may favor industrial organic over localist organic. Battles fought in one generation to gain public funding for organic research may have to be fought again in a subsequent generation, when new political administrations target organic farming for budget cuts or shift funding within organic toward programs more closely aligned with the needs of industrial organic farms.[16]

One should also keep in mind that the object conflicts in the scientific field extend beyond the issue of funding for organic agriculture research; they can also involve research agendas on the health benefits of food and nutrients. Also known as functional food research, the field is divided between an orientation that focuses on specific nutrients and their health benefits (which can then be added to processed food products) and an orientation that focuses on the health benefits of whole foods (for which specific nutrients are black-boxed or may be unknown). For example, a research agenda can be tilted toward documenting higher levels of a specific nutrient (such as omega-3 fatty acids that are found in grass-fed organic meats) or tracking a general health indicator for a black-boxed whole food, such as weekly consumption of grass-fed, organic meats in human subjects or animal models. To the extent that research agendas on organic foods tend to focus on the health benefits of specific nutrients (or the risks of specific pesticides or additives), they will tend to promote an understanding of "organic" as a sub-food entity (akin to a nutrient) rather than a whole food entity. Those agendas will in turn tend to favor an industrialized vision of organic as processed food rather than fresh, whole food, an understanding that in turn has implications for the localist aspirations of the sustainable agriculture movement.[17]

A second type of object conflict appears in the differentiation of food categories related to the consumer's decision to purchase organic or a related type of food. The food-processing industry has capitalized on general health consciousness and environmental awareness that favored organic food by developing the marketing categories of "natural" and "health" food, as well as food that is free from a substance that is perceived as risky, such as antibiotics or bovine growth hormone. However, categories such as "natural" or "health" food are unregulated and generally only have a vague meaning in terms of differential food quality. For example, "health foods" may be defined by the absence of partially hydrogenated oils, growth hormones, or some types of preservatives. They may use substantial levels of sugar, but not in a refined form, and consequently still pose substantial glycemic risk. Categories such as "natural" and "health" foods displace consumer attention away from organic products by diversifying the product choices. Health concerns are then divided against themselves, for example, in the choice between the nutritional benefits of a particular food choice versus the safety benefits of the claimed levels of lower pesticides or other contaminants in organic foods. Likewise, health concerns oriented toward nutritional benefits are separated from environmental and localist concerns associated with buying organic. The development of complementary categories—such as pesticide free, natural, and or the ubiquitous "health" food—therefore creates a broader, confusing field of healthy or green food options for which "organic" is diminished to the status of just one consumer choice among many.[18]

The third type of object conflict, product labeling and production standards, has received the most attention in the literature. The tensions between the social-movement side of organic food and agriculture and the increasing prominence of the industrial side became evident in a controversy during the late 1990s over the national organic food standard. In 1998 the Department of Agriculture proposed new organic food standards that would have allowed sludge, irradiation, and genetically modified seeds to be included in the definition of organic, and it would have also increased paperwork and fees for small farms. The Organic Consumers Association was founded in the wake of the threat, and it launched the SOS (Safeguard Organic Standards) campaign, which mobilized consumers through health food stores, farms, farmers' markets, and food coops. After the more-or-less successful campaign, the organization announced a much broader series of goals that included conversion of American agriculture to 30 percent organic by 2010, the phase-out of the worst industrial agriculture practices, and a moratorium on genetically engineered food and crops. The platform directly connected the pro-organic movement to the organizations involved in campaigns to label or limit genetically modified foods, such as Greenpeace and the Campaign to Label Genetically Engineered Foods. Meanwhile, in 1999 the standards-setting harmonization process continued at an international level via the work of the Codex Alimentarius Commission of the World Health Organization. Another standards battle emerged in 2005 and 2006, when an amendment was added to an agriculture bill to allow synthetic ingredients in the non-organic portion of food sold under the organic label.[19]

Even though the final standards of the 1998 controversy restored some of the crucial aspects of organic food quality, they created a new interstitial category of food label for processed foods, called "made with organic ingredients." In other words, the standards fragmented the object "organic" into a new set of categories of completely organic versus partly organic. The distinction may be defended as a helpful guide to consumers in the complex world of processed foods, but it also implicitly encouraged consumers to think of organic as a separate object from fresh, whole foods that have been grown on local farms and purchased directly from farmers. At the other extreme "organic plus" categories were emerging, such as biodynamic (a legacy of Rudolf Steiner), local organic, and fair-trade organic. Those categories can encode higher levels of sustainability in the production technologies as well as localist values such as preservation of regional farms. Consumers now face tradeoffs between labels that focus on the technical dimensions of the object as organic versus the societal dimensions of the object as contributing to locally owned businesses.[20]

Energy

As in the food and agricultural field, IOM organizations such as the anti-nuclear-energy organizations of the 1970s supported the development of TPM goals, such as the development of solar and wind energy. To build and maintain legitimacy, anti-nuclear organizations needed to propose a solution to the energy crisis rather than merely oppose the hazards and risks of nuclear energy and fossil fuels. Likewise, other environmental organizations also supported renewable energy, along with energy conservation, as a way of reducing health risks and greenhouse gases. Those efforts helped motivate and support a network of inventors and firms that was developing renewable energy technology.

Another important factor in developing TPMs for renewable energy was the support of both federal and state governments. In contrast with organic food and agriculture, the history of the TPMs in this field reveals the effects of earlier incorporation into government policy. In response to the oil crisis, in 1974 Congress passed the Energy Research and Development Act, which included a funding mandate for renewable energy research. The form and priorities of government support had an immediate effect on the prospects of the solar energy TPM. Two types of small-business entrepreneurs, one from the 1960s counterculture interested in the democratic potential of solar energy and another from small firms with an entrepreneurial interest, worked in 1975 and 1976 to develop proposals. Notwithstanding their differences in approach, neither of the two groups received significant funding, and to some extent the emergent TPM with its internal tension between social-movement goals and profitability goals was short-circuited. Both groups found that federal government funding instead went to large industrial corporations, which in 1974 had launched the Solar Energy Industry Association. The specter of homes and businesses abandoning the grid and of power lines coming down had been threatening enough to the electric power industry that it had launched its own version of renewable energy research. The crucial difference between the industry's vision of renewable energy and that of many of the entrepreneurs was that the industry had developed an on-grid design. The independent solar energy companies found that in order to get funding, they had to be associated with large corporations, and consequently there was a rapid phase-out of the small companies. Whereas the incorporation and transformation process took much longer in the food and agricultural field, in the field of solar energy it occurred much earlier as a result of the 1970s' energy crisis and the federal government's policy response.[21]

As the small companies were phased out or absorbed into the power industry, the technological field shifted away from alternative technologies and products based on designs that emphasized small-scale, off-grid independence for homes and businesses. Here, the complementarization process involved a shift from a locally controlled, off-grid vision of energy toward an on-grid vision that retained the power industry's control over renewable energy and captured the so-called free energy of the solar commons. Once the design of renewable energy was redefined in a way that was encompassed by the grid, it became acceptable to the power industry, which nevertheless did not invest enthusiastically in it. For the grassroots solar activists and some of the solar energy entrepreneurs, the industry's response represented an attempt to derail their vision of small-scale, locally controlled solar energy production. The relationship to the grid became one of the central conflicts for renewable energy, from the utopian visions of a decentralized, democratic solar economy of the 1970s to debates that emerged 30 years later about how to configure the planned transformation to the hydrogen economy. Although the social-movement approach to renewable energy did not disappear, and in fact it continued to grow through the home-power movement and the development of distributed generation, the approach became dissociated from the greening of on-grid renewable energy that the power industry was developing.

The effect of the early incorporation and transformation process can be seen in some of the organizational histories of this period, where social-change goals and mainstream industrial goals eventually collided. In 1978 Denis Hayes, a co-organizer of Earth Day, launched the first Sun Day and the first annual National Solar Congress, which brought together solar activists interested in a grassroots and small-scale approach to solar energy. Hayes's *Blueprint for a Solar America*, published in the following year, outlined a path for the United States to achieve 25 percent solar energy by 2000. In a speech on the first Sun Day, President Carter announced that he would support a new solar energy program with an increase in the budget to $100 million. Activists who had coalesced around Sun Day joined with business groups to form the Solar Lobby and Center for Renewable Resources. However, as the research program developed, proposals for community-oriented solar energy and small-business development were again dropped, and funding again went to solar energy proposals from large corporations. Only a small percentage of the budget went to appropriate technology, passive solar designs, and other small-scale, grassroots solar energy proposals. The Solar Lobby became torn between an activist side and a business side. After Ronald Reagan was elected president in 1980, the business side came to dominate the short-lived organization. The grassroots side survived at a regional level in organizations such as the Solar Center and Real Goods, both in California.[22]

The backlash against renewable energy was growing even before the election of President Reagan. Some of the federal and state tax credit programs for solar energy were criticized as too generous, and many of the first installations were subject to technical problems. The failures and the glitches could have been handled through reforms in the tax credit legislation and certification of installers. But once the Reagan presidency began, the federal government drastically cut solar energy support, and the emergent solar industry collapsed. The industry did not revive until the 1990s, when it underwent a growth spurt, and it received another boost in 2001, when California faced power outages in the wake of deregulation and energy-price manipulation.[23]

Wind energy, the other major exemplar of a new renewable energy source that does not involve generating additional greenhouse gases during electricity production, had a parallel history with some significant differences. Unlike in Denmark, where the development of wind energy technology can be traced to a prior grassroots social movement that underwent industrialization in a historical pattern that closely approximates the ideal typical TPM, in the United States wind energy had a longer industrial history and weaker social-movement participation. The use of wind power for water pumping was widespread in California's San Joaquin Valley during the late nineteenth century, and by 1889 there were 77 windmill factories in the United States. Although fossil-fuel-based pumps later displaced many of the windmills in California, wind power continued to be used to support water pumps on farms, especially on the Great Plains.[24]

After the 1970s, wind energy also became commercially viable at a larger scale than solar, and it became more attractive for grid-based energy production. Unlike solar energy, which had a more typical TPM trajectory of activists who advocated and in some cases took up entrepreneurial activity, wind energy in the United States had a less pronounced social-movement side. It is true that the original founders of the American Wind Energy Association, which began in 1974, have been described as hippies, and some of the people in wind power companies have been described as "ponytails" and even "Governor Moonbeam's children." But in the wind industry the countercultural element was only one strand among others, including entrepreneurs, corporate energy firms, former military engineers, and Wall Street investors. The American Wind Energy Association quickly developed a close relationship with the federal Department of Energy, and it eventually became the leading industrial trade organization for wind energy in the United States.[25]

This is not to say that there was no social-movement involvement in the development of the wind industry, particularly after the 1960s. In addition to advocates who entered the industry as designers and entrepreneurs, environmental organizations also pushed for green pricing programs and renewable portfolio standards. In the late 1990s, when green pricing programs were becoming widely available to electricity consumers in both investor-owned and public power utilities, advocacy organizations sometimes became involved in campaigns to encourage firms and households to shift to green power. In turn, the new strategy of grassroots campaigns to support signing up for green pricing schemes created divisions within the environmental community, because some organizations wanted to focus scarce resources on federal and state energy policies, such as the development of renewable portfolio standards. However, the broader point is that the debate between two strategies indicates that environmental organizations played a role in the development of wind energy in the United States during the late twentieth century, even though their advocacy presupposed and aided, rather than helped to generate, a wind industry.[26]

Like solar energy, wind energy benefited from government support programs, but the programs had a different effect on technology design and the technological field. Whereas even today solar energy has a largely modular design that allows some flexibility in scale and type of use, federal funds from the Department of Energy supported the development of large turbines oriented toward grid production and developed by corporations such as General Electric. During the period from 1974 to 1990 about $450 million of federal funding went toward large turbine development and only $50 million went to fund research on smaller machines, an example not only of the creation of undone science but also of the shifting of a technological field. The focus on developing large turbines also represented a complementarization of wind energy, that is, the transformation of its design to make it compatible with grid-based, centrally controlled generation and distribution. One example was the design of a large wind turbine by William Heronemus, who had left the navy and entered the wind industry over concerns with the hype around the peaceful atom and nuclear energy. Heronemus wanted his large turbine to be connected to off-grid hydrogen production, but General Electric and Lockheed, which became interested in the technology, preferred a configuration that involved sales to the grid. Similar priority setting occurred at the state level, although the wind energy leadership under Governor Jerry Brown tended to be somewhat more supportive of small turbine development. Together, the federal and state incentives triggered the grid-oriented wind rush of the early to mid 1980s.[27]

The split nature of American turbine designs (large turbines oriented toward the grid that did not work well and small ones for the home-power movement that did work) left open a space for the strong, efficient, mid-size Danish turbines. Marketplace domination by the more reliable machines of the Danish companies dashed hopes of creating a new industry and, with it, a political and economic base for further policies in support of renewable energy research and development. The lack of a significant domestic industry meant that there were few countervailing pressures to counteract growing popular skepticism, which grew due to a variety of factors: visible wind farms located near highways that were sometimes standing still, reports of bird deaths from the early wind farms that had been located in flyways, complaints about visual pollution in scenic areas, and failures of the some of the equipment. Almost all of the concerns were addressed in subsequent generations of wind turbine design, but the nascent industry had a long-term public relations problem. During the Reagan presidency, and especially after the decline of oil prices in the mid 1980s, government support for wind energy declined and the wind rush came to an end.[28]

By the late 1990s wind developers had solved the technical problems of large turbines, and the price of wind energy had become commercially competitive. The deregulation of electricity production markets and growth of state-level renewable energy portfolio standards created new opportunities for wind, which became the leading new source of "green" energy, an increasingly common category that was an indicator of emergent object conflicts. As with solar, the power outages of the late 1990s in California created a new political and economic opportunity, at least in that region. In wake of the outages, wind turbine use in rural areas of California grew rapidly, and one microturbine manufacturer claimed that business grew threefold in just one year alone, 2002, with more than half of the business in California. However, the growth was partly dependent on a federal tax credit, for which renewal was irregular and stalled in some years.[29]

This brief survey of the history of the solar and wind TPMs in the United States during the late twentieth century suggests some of the potential for variation in the history of TPMs, both within an industrial field and across them. One significant difference is the presence of early government support (that is, during the 1970s), which to some extent short-circuited the development of a reform movement of entrepreneurs by directing research funding toward grid-oriented, corporate developers. It was only with the development of technologies and policies in support of distributed energy with grid sellback that the trend toward large-scale, grid-based solar and wind energy shifted back somewhat toward the aspirations of grassroots control of the early solar energy activists. However, the new design of the object, distributed renewable generation, was far from the off-grid independence that one can find in some of the earlier visions of a solar economy.

As the industry and associated movement activity developed, one can again delineate various object conflicts. Conflicts around scientific research agendas were quite striking during the 1970s. The fundamental object conflicts for the solar research agendas took place around funding for large corporations versus entrepreneurial firms, with many of the small firms seeking to develop designs that were potentially grid independent and small in scale. Thirty years later the effects of the funding patterns were still felt, even as the opposition between on-grid and off-grid technology has tended to fade into the amorphous middle ground of distributed generation with grid sellback. Given the orientation of the wind industry toward grid-based production, the large turbines kept expanding in scale, but the small wind turbine market remained relatively undeveloped. Although there is an industry of microturbine manufacturers, as of the early 2000s few turbines with outputs between 10 and 225 kilowatts were available. There was some government funding for research into small-scale or appropriate technology (funneled through the National Center for Appropriate Technology), but the organization's budget for energy-oriented research was more focused on weatherization.[30]

A second type of object conflict within research funding has been over wind and solar together versus other forms of alternative energy development. Government-university-industry partnerships have increasingly focused on "clean energy" in the form of fuel cell development and technologies for controlling greenhouse gases from fossil-fuel sources, such as carbon sequestration. Funding from the Department of Energy tends to skew away from any kind of renewable energy, whereas the category of "clean energy" includes funding for research on "clean coal." Within renewable energy funding per se, a shift in priorities toward hydrogen has been achieved by failing to increase overall budgets for renewable energy research, therefore leading to reductions in available funding for wind and solar.[31]

Green pricing programs provided another site for object conflicts: the consumer decision. By the 2000s, many utilities offered a range of green pricing options and suppliers, so that consumers could decide what type of renewable energy source they preferred, usually at a price premium. For example, as of 2005 one major investor-owned utility offered four different renewable energy provider options: two were 100 percent wind power, one was 75 percent biomass and 25 percent hydropower, and one was 30 percent biomass, 40 percent wind, and 30 percent small hydro. At the time of writing, most green pricing programs across the country charged a premium for the average household of several dollars per month or about a penny per kilowatt-hour. By viewing the websites of the different providers, consumers could potentially also make choices among companies based on ownership and responsibility issues, such as private family ownership versus public ownership, local (within state) versus regional or national ownership, and companies that accept the principles of the Coalition of Environmentally Responsible Economies and those that do not. However, the rate of voluntary participation in green pricing programs has tended not to exceed 7 percent for most utilities.[32]

Regulatory standards for renewable energy have not been a major site of object conflicts in the energy field, partly because there is no national renewable energy portfolio standard. (As of 2005, there was a national renewable fuel standard, so one could examine object conflicts over definitions of what counts as a renewable fuel.) For the roughly 50 percent of states that had enacted renewable energy portfolio standards by 2005, the category of green electricity has been defined loosely to include hydropower and biomass. The definitional issues can be misleading, because states that appear to be on target to achieve relatively high levels of renewable energy in fact achieve those levels from already existing large hydropower, which can have severe effects on local ecosystems and human communities and can even be a significant source of greenhouse-gas generation. Another group of object conflicts involves standards used by utility companies to allow grid connection. The standards have tended to favor large producers and be very restrictive for small producers.[33]

A fourth type of object conflict has emerged for renewable energy. To date there appears to be no equivalent with organic food. (The closest equivalent might be concern with odors associated with suburban organic farms, but concerns with odors and other forms of pollution are much greater for factory farms.) Grid-based wind farms have sometimes been planned for scenic areas near mountains and/or coastlines, and the emergence of wind farms has generated a new wave of protest politics. Conflicts have emerged over visual and noise pollution and the general issue of rights to a viewshed. Interestingly, the conflicts have tended not to emerge in cases where the scale of the wind farm is smaller and energy control rests in the hands of the affected community, that is, where the sociotechnical system design was closer to early Danish model of locally controlled cooperative wind production. Unlike solar, which has generated little or no opposition (aside from a few complaints about glare), wind energy also created new divisions among environmental organizations, especially between the preservationist/wilderness organizations and the organizations most concerned with industrial pollution.

For example, the Audubon Society has very different views from Public Citizen over the value of wind energy. The avian concerns have been handled by charting bird flyways and more carefully managing the siting of wind farms, changing the design of the wind tower to make it difficult for birds to build nests in them, and slowing the speed of blades. Here, increased scale presents a potential solution to a preservationist concern, even as it has tended to coincide with non-local control and ownership.[34]

Waste and Manufacturing

The command-and-control and market-oriented policy instruments of late-twentieth-century environmental legislation, as well as the increasingly sophisticated mobilizations of communities affected by industrial pollution, created incentives for firms to consider how production and products could be redesigned. Although in some cases industrial polluters attempted to reverse existing regulations and resist new ones, they also invested funds into research and development in search of alternative technologies that complied with regulations at a minimal cost or even a profit. A regulatory push in support of the greening of industry coincided with increasing recognition of the potential of the profitability pull. The search began for design innovations that could allow firms to have their cake of regulatory compliance and eat new profits as well.

As industry began to face the upstream challenge of the redesign of products, their inputs, and production processes, a segment of the environmental movement began to operate downstream to intervene in the disposal of final consumer products. Often rejected (in many ways, justifiably) as middle class, ineffective, idealistic, and distracting, the history of the post-1960s recycling movement in the United States reveals a complicated political trajectory that at some points converges with, and at other points diverges from, the IOMs that emerged around industrial pollution. Ultimately, if one follows out the history into the 1980s and the 1990s, the threads of the recycling movement connect not only with the environmental justice movement but also with the greening of industry.[35]

Just as the post-1960s innovations in the fields of organic food production and wind energy drew on systems from a technological field that had been more prominent in the nineteenth century and the early years of the twentieth, so the emergence of the recycling movement as an expression of the 1970s mainstream environmental movement drew upon the longer history of private-sector and public-sector recycling. In the early decades of the twentieth century American cities were home to substantial recycling of solid waste that was carried out organizationally through the work of small businesses. Bottles were returned to local producers in a process that could today be called a "take-back" program, used textiles were collected to make rags, papers and scrap metal were given to scrap vendors, and organic wastes were used in gardens or returned to farms. At one point a substantial amount of food waste was recycled as hog feed, but the practice was discontinued when health researchers found that the practice was a risk factor in trichinosis. By the 1970s, scrap-metal collection and newspaper drives were still in use as a form of recycling, but other practices had subsided. During the post-World War II era producers had increasingly shifted to disposable products, and the longer supply chains that were a harbinger of late-twentieth-century globalization had disrupted local take-back relationships.[36]

Community-based recycling was occurring prior to the first Earth Day in 1970, but during the six months following the event an estimated 3,000 non-profit recycling centers opened across the country. The organizations were largely voluntary efforts that fed material to local scrap dealers, who in turn sold materials to remanufacturers, and consequently recycling during the period had the characteristic dual organizational structure of TPMs, where non-profit advocacy organizations were linked to private-sector firms. However, this particular TPM was very short-lived; from the existing sources it appears that all but a few of the post-Earth Day organizations did not survive the 1974 recession, when brokers were forced to slash prices. Although there was a sharp decline in the early 1970s, there was a second wave after the Resource Conservation and Recovery Act of 1976 mandated the closure of unsanitary landfills. As landfill closures mounted, local governments faced increasing pressure to find new solutions for solid waste, and some communities turned to recycling. In 1977 the California Resource Recovery Association was founded, and similar organizations followed in other states, so that by 1980 about 400 recycling organizations met at the First National Recycling Conference. Recycling continued to grow during the 1980s, and by the end of that decade there were about 5,000 organizations.[37]

During the 1970s and the 1980s, recycling organizations were mostly small businesses and non-profit organizations linked to local governments, but there were also some curbside programs run by city governments. The recycling organizations often combined environmental goals with community-development goals, and as a result local control and sometimes low-income employment and job training coexisted alongside the goal of waste reform. However, as David Pellow has noted, the otherwise laudable goals of some of the recycling organizations often had an unpleasant underside, because concern with community development and jobs for the unemployed coexisted with practices that exposed the workers to significant workplace hazards. Furthermore, the community-oriented aspirations were limited by the industrial partnerships with a largely non-local remanufacturing industry, which developed new products and technologies from recycled glass, paper, plastic, rubber, compost, and metal.[38]

Although there were possibilities for locally based remanufacturing firms, and in some cases recycling did help to launch local remanufacturing businesses, in general the remanufacturing side of the TPM did not take the form of small, locally owned business development. As in other manufacturing industries, remanufacturing firms that increased scale and sought out larger markets were better able to maintain a steady, low-cost supply of inputs as well as stable, increasing demand for the products at competitive prices. In some cases recycled goods could be absorbed by the local economy, such as the use of crushed glass in roadbeds or recycled glass for artistic glassworks, but in general the recycling of the waste stream tended to be a non-local enterprise and even global in scope. Because recycled materials are inputs into a remanufacturing process, it is difficult for small manufacturers to compete in local markets when similar goods are coming in from outside markets with lower prices due to economies of scale. Furthermore, remanufacturing based on locally obtained recyclables is highly reliant on local supply chains and therefore vulnerable to the greening of production practices that recycling and zero-waste advocates ultimately want to see upstream. In this sense, whatever the community-development aspirations of some non-profit recycling centers, they were tied to a non-local remanufacturing industry. Still, the private-sector symbiosis was crucial to the development of the TPM, because the voluntary recycling organizations could not exist without a market for recycled goods.

A second major problem that recycling encountered, especially during the 1980s, was competition with another growing form of waste management: incineration. In 1969 Keep America Beautiful, an industry NGO formed by packaging and product companies to divert attention from packaging waste to litter, spun off an organization that became the National Center for Resource Recovery, which advocated incineration as a strategy for solving solid-waste excesses and energy shortages. Although the number of incinerators declined during the 1970s from about 369 in 1969 to 67 in 1979, during the 1980s they underwent rapid growth, with 110 in operation and another 200 in the planning or construction stages by 1987. An important factor in the turnaround for the incinerator industry was the passage of the Public Utilities Regulatory Policies Act of 1978. The law altered economic conditions by making it possible for waste-to-energy facilities to sell electricity to power companies at favorable rates. In 1979 the Department of Energy and the Environmental Protection Agency endorsed incineration as the preferred waste disposal technology, and the Reagan administration encouraged the privatization of solid-waste management. Another factor in favor of incineration came from the ailing nuclear energy industry. As the plans for new nuclear energy facilities wound down during the late 1970s, the manufacturers of nuclear energy reactors shifted to incinerator production.[39]

Recycling advocates were divided over incineration, and here we begin to encounter the central issue of the incorporation and transformation of the recycling TPM. One perspective was articulated by the National Recycling Coalition, which was founded in 1978. The organization redefined recycling in two fundamental ways that were consistent with two different industry interests: recycling as a complement to other forms of waste disposal, including incineration; and recycling as a for-profit, curbside project that could be integrated into the waste management industry. At the local level some recycling organizations lined up with incineration, and their support put them in opposition to environmental justice groups and local anti-incineration coalitions. In contrast, some recycling activists continued to push for a form of recycling that emphasized non-profit organizations, local economic development, and opposition to incinerators. They operated largely through state-level organizations such as the California Resource Recovery Association and through national NGOs such as the Institute for Local Self-Reliance. The anti-incinerator side of the recycling movement converged with the emerging environmental justice movement, and some of the recycling activists played a role in growing local opposition to incinerators. By the late 1980s public opposition to incineration and mounting costs from increased requirements for pollution control technology had, in ways that bear comparison with the downfall of nuclear energy during the 1970s, led to the demise of incinerators. Meanwhile, partly to divide and weaken opposition to anti-landfill and anti-incinerator coalitions, waste management firms and incinerator supporters began to support an integrated strategy that included recycling facilities.[40]

Although opposition to incineration climbed through the late 1980s, landfills were also under pressure to close, expand, or modernize. The waste crisis achieved high levels of media attention with stories such as the Mobro 4000 garbage barge, which in 1987 went on its historic 6,000-mile voyage from port to port in search of a landfill. As a result of public opposition to incineration and, in many cases, also to new landfills and landfill expansion, state and local governments began to support recycling in a much more concerted manner, and by 1989 the Environmental Protection Agency began to shift away from the incinerator model in favor of recycling. By the late 1980s, the waste management industry had also undergone considerable consolidation and vertical integration, and consequently it was in a better position to step in and provide a version of recycling as an alternative to increased incineration. The waste industry's version of recycling incorporated grassroots recycling by transforming the design of programs from non-profit, locally controlled, community-development goals to curbside pickup that was integrated into the waste industry.[41]

As recycling became mainstream, governments at various levels initiated procurement policies to develop demand for recycled products, and the 1990s became a decade of explosive growth in recycling. Curbside recycling programs grew tenfold during the decade to reach about 10,000 in the year 2000, and by that year approximately 33 percent of the nation's waste stream was diverted into recycling. More generally, by 2000 the reuse and recycling industry had grown to 56,000 establishments with about a million jobs and $236 billion in annual revenues. As city governments contracted with waste management firms to handle curbside recycling programs, the non-profit, community-oriented recycling organizations that had survived from the 1970s and the early 1980s were marginalized, put out of business, or forced to redefine their model from recycling to reuse.[42]

Although the statistics on recycling seem impressive, they can be placed in an alternative light that reveals as many shortcomings as successes. As David Pellow, Allan Schnaiberg, and Adam Weinberg have documented, the shift in recycling from community-based, non-profit organizations to large corporate firms in the waste management industry has resulted in higher levels of workplace hazards for workers in recovery facilities, dependence of regional governments and economies on large corporations, and the dumping of toxic waste in foreign countries. One would think that with all of the shortcomings there might be a silver lining of increased efficiency in the form of high recovery rates, but the recovery rate for recycled material in the United States remained well below that of other countries. By the early 2000s some cities in California had achieved a state target of 50 percent, and a few were on their way toward a goal of zero waste, but the phase-out of landfilling as a form of solid-waste disposal remained an elusive goal for most American cities. American cities were generally recycling at rates well below some of the major Canadian cities, which had achieved a recycling rate of up to 70 percent. Furthermore, many of the "recycled" products were only down-cycled (for example, newspapers were often recycled as animal bedding), and consequently recycling came to mean a short-term delay in the trip to the landfill or incinerator. Even if the rate of recycling and the growth of downcycling could be fixed, the general ecological problem of increased consumption remained unsolved: recycling has not reduced the growth in the absolute amount of waste. In 1990, 117.5 tons of packaging and throw-away products went to American landfills. By 2000, even with the growth of recycling programs, that had grown to 121.3 million tons. Not only did Americans generate approximately four times as much waste as Japanese and Danes; they generate more and more waste.[43]

The option pursued by activists of organic farming and renewable energy, to regroup around the politics of local control and community development, has not been as available in the recycling movement. To some extent activists failed by not developing partnerships with locally oriented remanufacturing firms, but, as I argued above, such partnerships would have likely been difficult to achieve given the benefits of non-local production sources and markets in this industry. Although there are economies of scale in both agriculture and manufacturing, the premium that consumers are willing to pay for fresh, local produce with a known provenance does not translate as easily for products based on locally recycled paper, plastics, glass, and metal. However, the grassroots recycling activists did reconstitute in two ways: merging into a new, more comprehensive solid-waste movement that showed increasing convergences with the environmental justice movement, and shifting to a more localist variant in the form of reuse centers. The first variant will be considered here, and the second in the next chapter.

Some zero-waste activists and advocates in the United States claim that their work grew out of long-term activist struggles to stop landfill and incinerator projects. Zero-waste activists resituate recycling, composting, and reuse in broader campaigns aimed at reducing packaging, non-recyclability, and the toxicity of manufactured products. Zero-waste action pursues a dual strategy that is aimed at reform of policies and practices of both community governments and firms. A central concept behind zero-waste action is "extended producer responsibility," which requires manufacturers to redesign products to reduce content that ends up as solid waste and to mandate producer "take-back" of used products. The goal is to shift responsibility back to producers from the shoulders of local communities and end users, which not only must pay for recycling or landfill costs but also must endure the toxic effects of waste processing and disposal. In the United States the politics of zero waste led to a division within the National Recycling Coalition in 1996, when zero-waste advocates left to form the GrassRoots Recycling Network. The latter organization has supported both community-oriented zero-waste policies and extended producer responsibility.[44]

Because extended producer responsibility campaigns aim to force manufacturers to take back products, successful campaigns will motivate manufacturers to redesign products to make them more recyclable and therefore less toxic. Although there are many legislative developments in the European Union and in wealthy countries such as Japan, in the

United States the more neoliberal orientation of Democrats and Republicans alike has protected industries from producer responsibility laws, and there has been little action at the federal level. Given the lack of leadership at the federal government level, state governments and environmental organizations have often filled the gap. For example, the governments of Iowa, Minnesota, and Wisconsin, together with some support from the Environmental Protection Agency and NGOs, developed an agreement with the carpet industry to develop take-back goals and programs. Zero-waste activists have also helped local governments to set dates for the closure of incinerators and landfills, ban all organic material and recyclables from landfills, and place a surcharge on landfilled waste. Policies that block standard landfill options are coupled with incentives for recycling such as garbage lotteries and programs to help set up reuse and repair centers. In turn, the policies help reinvigorate the older reuse and thrift industries of the local economy.[45]

Activists have also targeted corporations to develop take-back and zero-waste programs. One example is the Computer Take Back Campaign developed by a variety of environmental and environmental justice organizations, including the GrassRoots Recycling Network. The campaign included legislative initiatives in twelve states and targeted Dell Computer, which had the advantage of extensive customer records. The campaign's goals include producer take-back, environmentally superior recycling, phase-out of hazardous materials, no prison labor, and no export of hazardous waste. Likewise, the Rainforest Action Network (better known for its campaigns against the use of tropical hardwoods) launched its Zero Emissions Campaign and targeted Ford Motor Company for its low fleet-wide fuel economy average.[46]

The Computer Take-Back and Zero Emissions campaigns focus on ending specific practices and therefore engage in the politics of the moratorium that is continuous with IOMs, but they also encourage upstream transformations of product design so that toxic waste generation and wasteful fuel efficiency do not occur in the first place. In other words, by forcing products to be returned to factories, producers are motivated to replace toxic materials with less toxic substitutes, as is occurring in computer manufacturing, and as a result take-back programs create pressure on meaningful product redesign. Ultimately, then, one positive legacy of the recycling movement of the 1970s, at least the post-1980s transition of a portion of it as a contributory stream in an emerging zero-waste movement, is to create pressure for the redesign of products and production processes.

In summary, the recycling movement has undergone various transformations that have contributed to and coincided with an altered technological field. At first operating outside the official industry and local government programs of waste management, recycling was eventually incorporated into a consolidating and differentiating waste management industry and has become a significant feature of the technological field of the industry. In the process the design of recycling programs has been transformed from small-scale and non-profit drop-off programs to corporate and for-profit curbside programs that separate materials in large recovery facilities for sale on global markets. Some of the surviving recycling centers were driven into the reuse industry, and some recycling activists shifted to the upstream politics of producer responsibility. The zero-waste activists have placed pressure on firms to develop the ecological redesign of manufacturing processes and products, especially around product take-back policies and programs.

As the politics of redesign develop, object conflicts related to recycling and remanufacturing can again be tracked in three major sites. Research fields may be in the process of shifting toward sustainable design, but evidence for such reorientation in the United States in fields such as industrial design is still scanty. Although there are leaders such as William McDonough, in the United States sustainability issues do not have the same salience to date that they have in parallel fields in design and engineering in Europe. Green chemistry appears to be gaining some ground, and the rush toward nanotechnological materials may provide some greener substitutions, but the first wave of research on nanotechnological materials suggest that some materials, such as carbon-based buckyballs, may present a new level of health and environmental risks.[47]

Regarding consumer decisions, a niche market has emerged for consumer products with recycled content. Object conflicts can emerge over the degree of recycled content and the relative safety or toxicity of recycled material. For example, procurement policies in favor of purchasing materials with recycled content or recyclable content face definitional issues over the percentage of the product that must be made with recycled content and what kinds of products should be targeted for procurement preferences. During the 1990s state and local governments set procurement standards for recycled content, especially for paper, and in 1993 an executive order by President Clinton established a goal of 30 percent recycled content for federal government paper. The standards raise many definitional issues, such as what are optimal percentages in terms of product quality, price premia, and supply availability and which products are targeted for recycled content.[48]

Regarding regulatory policy, object conflicts over what extended producer responsibility or take-back means are also underway. In the United States the goal of developing producer responsibility laws and ubiquitous take-back programs has been met with the counter-proposal of "product stewardship." Unlike extended producer responsibility, product stewardship would replace the mandatory government regulations found in Europe with voluntary compliance and would disperse responsibility from producers to users. Regarding take-back programs, voluntary programs have been instituted as a strategy to avoid legislative intervention. For example, in 1994 several major rechargeable battery firms formed the Rechargeable Battery Recycling Corporation in response to the threat of impending state regulations, and a subsequent industry-based electronics program for recycling in California was inaugurated in response to a threat from the California Senate to enact legislation for mandatory recycling. Participation rates in the voluntary programs have been generally low; for example, a study published in 2003 concluded that one large computer manufacturer sold only 200 contracts per month for its computer recycling program, and recycling rates for rechargeable batteries fell far short of targets. In another case, a computer firm backed take-back legislation only after bad publicity emerged over the recycling of its products to China. In the United States to date extended producer responsibility has a history of being mostly public relations: to date, EPR is mostly PR.[49]

Infrastructure

The late-twentieth-century efforts to reform urban design and building design formed a web, with no single central TPM. I will focus here on reform movements within the planning and architecture professions. The movements operated within a complex institutional field that included, on the one side, coalitions with grassroots social-movement organizations around issues such as tenants' rights and transit access, and, on the other side, the real estate development and construction industries as well as local, state, and federal governments.

The emergence of the profession of planning was itself a response to urban congestion and other urban problems, and consequently the profession has always had an ameliorist orientation that can be found across its many schools and strands. Throughout the profession's history in the twentieth century various political reform movements have also played an influential role. For example, the progressive and feminist movements of the early twentieth century influenced a strand of planners prior to World War I, and after World War I a democratic and environmental political vision was articulated in the Regional Planning Association of America and the work of Lewis Mumford. After World War II the critical tradition continued in the work not only of Mumford but also of Herbert Gans and Jane Jacobs, who saw the profession as subservient to development interests and questioned the benefits of housing projects and other forms of urban redevelopment. Although reformist aspirations were widespread, planners never achieved the level of autonomy found in other professions, such as medicine and law, and instead they often found themselves in weak institutional positions within urban political administrations that were ruled by growth coalitions.[50]

In 1964 reformist planners founded Planners for Equal Opportunity to identify the effects of planning on the poor and ethnic minority groups, and within a few years the American Institute of Planning had responded by recognizing the importance of issues of race and poverty. Although the alternative organization was disbanded, new organizations emerged in its wake. Urban Planning Aid, the Boston organization that had played a prominent role in the city's anti-highway movement, continued to evolve toward a social-movement organization dedicated to building a working-class coalition with goals of rent control and support of the struggles of tenants' unions. More generally, organizations such as Architects, Designers, and Planners for Social Responsibility; the Planners' Network; and the Association for Community Design provided a meeting ground between professional reform action and grassroots organizations that emerged from the civil rights and anti-poverty movements, including tenants' rights organizations.[51]

In the 1980s a more moderate reform movement emerged, and in 1993 it was institutionalized as the Congress for the New Urbanism. New urbanism involved both architects and planners, but unlike the post-World War II profession of planning, which had come to focus on urban policy, New Urbanism brought an architectural sensibility to issues of urban design and was seen by some planners as an architectural reform movement. Among the better-known principles of New Urbanist design were pedestrian-friendly neighborhoods, mixed-used buildings and zoning, transit-oriented development, public green spaces, and preservation of older architecture. The charter's goals suggested that New Urbanist projects would benefit low-income residents of cities through, for example, enhanced access to urban transit and mixed-income housing.[52]

However, there were some crucial differences between the New Urbanist visions of design and justice and those of the grassroots groups and radical planners. A significant difference was the tendency of the New Urbanist projects to filter out the participatory aspects of urban design. Although low-income residents would benefit from New Urbanist desiderata such as proximity to transit stops, pedestrian-oriented streetscapes, and mixed-use buildings, they were not necessarily brought into the decision-making processes around goal prioritization, and they could just as easily find themselves pushed out of neighborhoods that were undergoing gentrification. In contrast with the design priorities of the New Urbanists, radical planners used participatory methods and sought goals of rent control, bills of rights for tenants, and even zoning changes to allow home businesses.[53]

As professional reformers New Urbanists needed the real estate development industry to move their projects off the drawing boards, and consequently they formed an industrial partnership that is characteristic of TPMs. Here, the social-movement side of the TPM was divided between the radical planner-activists and the professional reform movement of New Urbanists, especially New Urbanist architects. Many of the first projects of New Urbanism were greenfields housing developments built for middle-class home buyers. The application of the reformist principles to suburban developments resulted in both the incorporation of reformist principles into the housing industry and their transformation. Showcase projects—such as Seaside in Florida, Kentlands in Maryland,

and Laguna West in California—were valuable as demonstrations of the economic feasibility of suburban developments based on pedestrian-oriented design with neotraditional town centers, but they also provided limited examples of the full New Urbanist reform vision.[54]

Neotraditionalism in greenfields housing developments became increasingly incorporated into the suburban development designs of the housing industry. According to the publisher of *New Urban News*, neotraditional developments across the United States grew from a small number in 1990 to 750 in 2005. The new developments offered front porches, sidewalks, facsimiles of town centers, and occasionally alleyways and walking paths as elements of a more walkable and community-oriented design. Although some of the crucial reforms associated with New Urbanist design (especially transit-oriented development and mixed-use, mixed-income buildings) were often deleted from the neotraditional design, the new developments represented an attractive symbolic expression of a pre-globalization community space of small communities in more localized economies. Popular with home buyers, the developments commanded a 20–30 percent price premium, even when the "town centers" offered little more than a few retail shops, a small post office, and an elementary school. In some cases residents even formed their own grassroots organizations to protect their neotraditional neighborhoods against proposed incursions by retail developers, such as a proposal for a strip mall instead of a traditional town center in Gaithersburg, Maryland.[55]

With the incorporation of New Urbanism into the real estate development industry through a design shift into suburban housing developments, the reform movement earned the unflattering sobriquet "the new suburbanism." Critics, including some prominent planners, charged that New Urbanists had turned their backs on urban poverty issues. In reply, New Urbanists could point to their influence on numerous projects that were constructed in urban areas and, more generally, to the growing diversity of their professional reform movement. During the Clinton years the federal Department of Housing and Urban Development embraced New Urbanist principles and funded projects through its HOPE VI Program, and by 1999 there were 155 New Urbanist infill and brownfields projects. Nevertheless, even the New Urbanist projects that were oriented toward urban redevelopment came under criticism from grassroots design and anti-poverty leaders, because some projects displaced current residents with vouchers and replaced renters with higher-income homeowners. According to one leader, Michael Pyatok, the conflict among design profession reformers was especially clear when both the Congress for the New Urbanism and the Association of Community Design Centers were planning to meet in Portland, Oregon, but the Congress for the New Urbanism refused to hold a joint panel discussion or even to advertise the existence of its poor cousin.[56]

Unfortunately, the association of New Urbanist infill projects with the risk of gentrification has played into the hands of "big box" retail developers. They have been able to rally neighborhood support by portraying New Urbanist visions of walkable neighborhoods with small shops as a gentrification process that ends up providing residents with boutiques for the middle class rather than access to affordable, practical retail such as supermarkets. By reframing urban "big box" developments as benefiting the urban poor rather than a gentrifying middle class, in some cases developers have been able to turn low-income urban residents against New Urbanist proposals.[57]

A second criticism of New Urbanism that is also relevant to the issue of incorporation and transformation focused less on the class and poverty concerns than on the shortcomings of suburban greenfields projects from a regional planning perspective. Because neotraditional suburban developments fail to implement connections with transit systems and link housing developments to a broader context of urban design and planning, they may end up promoting a slightly more dense form of urban sprawl. Although the criticism is valid up to a point, one New Urbanist leader, Peter Calthorpe, has focused on a larger scale that situated transit-oriented development within regional planning that is oriented toward the regional ecology and includes proposals for tax sharing across municipal boundaries. The shift in scale is one example of the articulation of New Urbanism with the policy concerns of urban planners who work on sprawl, green spaces, and the rationalization of urban growth.[58]

Planners have tended to work on similar goals under the banner of "smart growth." For example, in the mid 1990s the American Planning Association launched its "Growing Smart" initiative, which led to recommendations on model statutes. Some planners were originally critical of New Urbanism, but in 2000 the American Planning Association began the process of creating a New Urbanism Division, and by 2005 there was considerable confluence between New Urbanism and smart growth. The Smart Growth Network was funded by the Environmental Protection Agency in 1996, and by 2004 Smart Growth America claimed to have 100 advocacy organizations as members, including environmental, urban poverty, real estate, and governmental organizations as well as the Congress for the New Urbanism. Goals vary from organization to organization, but they include preservation of open spaces, development of walkable cities and public transportation, and infill instead of more green space development. Social equity issues are also included in the goal statements, but the primary orientation is land use policy such as green space preservation and the development of urban growth boundaries. The policies are particularly attractive to rapidly growing urban regions. The history of smart growth networks is not yet written, and I have not identified an incorporation and transformation process similar to that of the neotraditional suburban development for New Urbanism. However, one might argue that very concept of "smart growth" represents another incorporation of planning reformist agendas into the values of the traditional urban growth machines and the associated development industries.[59]

Another close cousin of New Urbanism, but in the opposite direction in terms of scale, is the reform movement that has emerged around green building design. Advocates trace their history back to the passive heating and cooling systems of nineteenth-century buildings such as the Crystal Palace in London, but the more proximate history in the United States began with the response to the energy crisis of the 1970s. Unlike some TPMs evidence for a social-movement component is difficult to find other than participation from environmental organizations in the development of green building codes and a grassroots tradition of green building associated with the home-power, straw-bale, and "back to the land" movements. However, for commercial green building codes the available history, which is not yet written professionally, focuses more on the response of the American Institute of Architects to the energy crisis of the 1970s, when the professional organization formed a Committee on Energy. During the late 1970s the Department of Energy and what later became the National Renewable Energy Laboratory also began addressing energy-related issues in buildings. As a result the TPM around commercial green building emerged from professional and governmental groups. In 1993 the architectural group joined with others to develop what became, in 1998, the LEED (Leadership in Energy and Environmental Design) standards for commercial buildings.[60]

It is worth dwelling for a moment on LEED standards, because there is some evidence for a process of standards modification that bears similarities to the emergence of neotraditional suburban development designs in the hands of the housing and real estate development industry. LEED standards were first adopted by federal, state, and local government agencies as well as by non-profit organizations, and it took more time for the standards to gain visibility in corporate real estate. Nevertheless, real estate under LEED certification grew rapidly from 1.1 million square feet in 1999 to an estimated 229 million square feet in 2005. Points are assigned based on site sustainability, water efficiency, energy and atmosphere, materials and resources, indoor environmental quality, and innovation and design processes. Levels of certification are based on the total number of points, from certified to silver, gold, and platinum levels. An analysis of the most popular criteria for points indicated that contractors tend to opt for the low-hanging fruit, in terms of cost and ease of incorporation into construction projects, and they shy away from dramatic energy improvements and renewable energy. LEED standards were also under development for residential buildings, for which standards were formerly constructed in a variable manner at the local level.[61]

Although LEED standards represent a development guided by a multi-organizational reform movement that works with the construction industry, there is some evidence that the residential segment of industry is responding with a different set of standards. In 2005 the National Association of Home Builders announced its own voluntary green home building guidelines. The extent to which those guidelines will water down and displace the LEED guidelines remains to be seen. Other standards are being developed in the opposite direction: a more rigorous alternative to LEED involves the "Systematic Evaluation and Assessment of Building Environmental Performance" to assess the conflicts and synergetic effects among the various green design features.[62]

To summarize, there is some evidence for the transformation of original visions of change, such as in the case of suburban neotraditional housing developments and the use of the term "growth" as a frame for environmentally oriented regional planning. A suggestion of a similar transformation is also evident in the history of green building standards, where again the housing industry's alternative standards may end up watering down LEED standards (which themselves have been accused of encouraging cherry picking). The general influence of the reform movements on the technological field of urban and suburban infrastructure is dubious. Although I have cited statistics that indicate high levels of growth on specific metrics such as neotraditional housing developments and square footage under LEED certification, the numbers are arguably dwarfed by the ongoing construction of superstore complexes and low-density, suburban housing developments. As in the cases of organic foods and renewable energy, we are encountering rapidly growing niche markets that are simultaneously undergoing dilution and differentiation as they scale up.

Object conflicts in this field of TPMs again can be examined in three major sites. Regarding conflicts over agendas for scientific research fields, there is very little information available, so I am limited to making a few suggestions that might be examined in future research. Research programs oriented toward smart growth appear to have a good position in the planning field and can be considered quite mainstream. To some degree they are continuous with the profession's long-standing reformist aspirations with respect to urban design in the public interest. In contrast, sustainable building design can be found but does not yet dominate the agendas of the country's major architecture programs, at least in the elite, private universities.[63]

Regarding conflicts over consumer purchases, builders of neotraditional neighborhoods or green buildings must be able to convince consumers that the alternatives are worth the price premium. New suburban developments based on some elements of neotraditional design have been able to command a price premium, but green housing has to date largely been translated into energy efficiency metrics, such as insulation and window quality. Given the association that middle-class buyers have between crime and the design features of mixed-income housing and public transit, it seems likely that market-based mechanisms will encourage continued watering down of the original, holistic vision of New Urbanists' and planners' goals of more extensive regional transit systems.

Object conflicts also occur over standards and codes that define what can and cannot be built. For example, zoning codes can be restrictive for builders who want to develop mixed-used, multiple-income neighborhoods. Likewise, building codes can restrict some green building innovations, such as on-site wastewater treatment. Although the design of LEED standards is itself a site for object conflicts, I have also suggested that green building standards appear to be in the process of becoming a site for more extensive object conflicts, as the construction industry and professional reformers develop alternative standards to LEED.[64]

Finance

TPMs in the financial field begin with a basic tactic of social movements, the boycott, and transform it into the "buycott," or targeted spending in favor of an alternative product. The development of the buycott for consumers may accentuate the positive (the desirable product due to an investment screen or product label), but it does not eliminate the negative politics of the boycott; rather, the buycott combines the two by shifting consumption away from some products and toward others. Two major TPMs have emerged in the financial field during the late twentieth century, one oriented toward investment and the other toward consumption. They are continuous in that both apply value-based screens and product labels to shift consumer preferences for products. Organizationally, however, they are more or less distinct, because one focuses on financial products and investments and the other on consumer product choices. My discussion of TPMs in the financial field will be briefer than the previous sections, partly because the evidence for the incorporation and transformation process is either questionable, as in the case of socially responsible mutual funds, or of much more recent nature and still in formation, as in the case of consumer product labels.

Socially responsible investment has a long history in the United States that is sometimes traced back to the Quakers' refusal to invest in slaves and weapons. The first mutual fund to engage in ethical screens was the Boston-based Puritan Fund, which has operated since the 1920s. Another precursor of ethical investing was the National Council of Churches' request that member denominations sell their holdings in companies that practiced racial discrimination. In 1971 the Pax World

Fund was founded, and in subsequent decades mutual funds with various types of ethical screens were offered. In 1974 the Interfaith Center on Corporate Responsibility was founded from groups within the National Council of Churches to provide coordination among faith-based investors. Religious institutional investors advocated divestiture from companies with investments in South Africa and weapons, and they developed coalitions with public pension funds as well. Alongside the growth of the industry, the student-based anti-apartheid divestiture movement of the 1970s and the 1980s spurred universities to develop investment screens. At an organizational level, socially responsible investment exhibits the TPM dynamics of a combination of NGO advocacy organizations, including faith-based organizations and student organizations, and private-sector investment firms that offer socially responsible investment products.[65]

The emergence of the industry of socially responsible investment was so closely related to the anti-apartheid movement that its leaders were concerned that the field would wane after 1993, when South African President Mandela requested that the divestiture movement be ended. Instead, socially responsible investing went through a period of rapid growth. Total assets grew from $40 billion in 1984 to $639 billion in 1995 and $2.3 trillion in 2001 (counting screened institutional investments), and the total number of socially responsible mutual funds grew from 55 in 1995 to 181 in 2001. As with organic food, renewable energy, neotraditional housing, and green buildings, the rapid growth should be contextualized against an overall small market share. Although the field was growing at a more rapid rate than the total investment field, assets in socially responsible investment amounted to only about 10 percent of the total investment funds in the United States.[66]

To focus on socially responsible investment in the mutual fund arena, there is a pattern of relatively small firms that were market innovators and leaders, such as Pax, Calvert, and Domini. Give the high growth rates, one would expect more established firms to enter the market, as occurred with large food-processing companies for organic food products. Indeed, some of the large mutual funds, including the Dreyfus Group and Smith Barney, also entered the field. However, the large funds appear not to have squeezed out or acquired the smaller ones (as occurred in the organic food-processing industry), perhaps because both the small and large funds had sales loads or relatively high expense ratios. But in 1999 the Dow Jones Sustainability Indexes were launched, and in 2005 indexed responsibility funds had assets of $3 billion. In 1999 the relatively small Calvert Group (about $10 billion in assets as of 2005) partnered with mutual fund behemoth Vanguard ($800 billion in assets in the same year) to create a social index that Vanguard could offer in its family of low-cost index funds. The new index fund, which became available in 2000, offered significantly lower fees and could displace some of the older funds. To the extent that one can locate a process of incorporation and transformation in the socially responsible investment field, it will be the transition from small, actively managed funds to large, index funds with their razor-thin expense ratios or even exchange-traded funds. But it is not likely that index funds will ever replace the smaller, actively managed funds. Not only do investors have different views about the economic value of index funds versus managed funds, but they also may assess the politics and values of the two differently.[67]

As the field of socially responsible investment funds grew, there was considerable product differentiation. Investment screens were originally rooted in the progressive social movements of the 1960s and the 1970s, but they have diversified and can include conservative and general moral values as well. Common screening criteria are tobacco, the environment, human rights, employment/equality, gambling, alcohol, and weapons. Community investing, community relations, and labor relations constitute a smaller percentage of screen criteria but are found in 30–40 percent of the funds. Some funds, such as Winslow Green Growth, use only environmental criteria, whereas others, such as Calvert New Vision, mix environmental, community, and equity criteria. In other words, at this point the field is diversified enough that it offers many flavors of the object of a "sustainable" investment.[68]

The second major TPM in this pathway involves consumer product labeling. Probably the most significant example of the buycott for retail consumer products is the alternative trade movement. This TPM dates back to the 1950s, when religious groups starting selling artisanal products from poor countries through their congregations. The organizations grew into Ten Thousand Villages, a network of stores developed by Mennonites, and a parallel network developed by the Brethren called SERRV International. Various non-religious organizations emerged during the 1970s and the 1980s, and in Europe three labels, mostly for coffee, were introduced in the late 1980s. Fair-trade groups began meeting in conferences, and in 1992 the North American groups formed an umbrella organization, the Fair Trade Federation. In turn, the international organization, Fairtrade Labeling Organizations International, was founded in 1997.[69]

Fair-trade labels assure consumers that suppliers are relatively disadvantaged and that independent monitoring systems are in place. Usually the farmers receive a price premium on their goods in addition to access to stable markets and prices. Fair-trade labeling diversified from coffee and cocoa to tea, sugar, fruits, and wine, and by 2002 world sales were at $400 million and growing at an annual rate of 30 percent. Again, in absolute terms the market remained tiny. For example, at that time fair-trade coffee accounted for only 0.8 percent of global coffee sales, in comparison with organic coffee, which accounted for only 0.58 percent of sales. A much smaller percent was double certified as "organic fair-trade."[70]

Even though fair-trade initiatives remain a tiny portion of commodity markets, they have had an effect on the major consumer retail firms. Here, we again see the phenomenon of a mixture of success and cooptation as the TPM is incorporated and transformed by an existing industry. Fair-trade retail sales in developed countries such as the United States began in independent nonprofit stores, but the products have since become available in corporate retail chain stores. The change has allowed the fair-trade TPM to scale up, but for advocates of "buy local" campaigns the change has meant that "fair trade" is only fair at the producer end of the commodity chain. Furthermore, large corporations such as Starbucks and Nestlé have developed their own producer preference systems, which give points for social responsibility as well as environmentally sound production and processing. As the corporate systems incorporate and transform the organic/fair-trade labels, they tend to delete out some of the features in the original systems, including price premia for producers, independent certification boards, and participation of farmers in setting standards. It remains to be seen how such systems develop and how beneficial they are to low-income farmers in developing countries.[71]

In the financial field, object conflicts focus on the consumer, including institutional consumers of financial products, more than research agendas and regulatory policies. Socially responsible investing has become differentiated to the point that there are many different types of investment screens on the market. The diversity of options raises fundamental questions about what an environmental screen means or how an investment in one publicly traded corporation can be considered more sustainable than in another one. Although a screen might preclude a company that has investments in fossil fuels, the fund may end being heavily weighted in technology stocks such as software companies. It is not clear how meaningful it is to consider such investments green or sustainable. For consumer commodity purchases such as coffee, object conflicts are emerging over which type of product label consumers will use as a guide in making decisions: a corporate sustainability label or the original, stronger fair-trade labels, which can be accompanied by double certification as organic. If the experience with timber certification discussed in the previous chapter is a guide to the future, then ongoing negotiations between NGOs and corporations will tend to discredit ersatz labels as they arise and push the corporations toward alignment with the goals behind the original standards.

Conclusions

In the case of TPMs relatively small networks of non-profit organizations, reformist professionals, and for-profit firms, often with low capitalization, have been able to develop rapidly growing markets for alternative products based on alternative production technologies. The bar for declaring a TPM successful can obviously be set at different heights, but even if the alternative technologies and products constitute a small absolute percentage when compared against a broader market volume, they generally exhibit rapid growth. Furthermore, success might be defined less in terms of market share and more in terms of the ability to leverage some innovation and change in the technologies and products of the target industries.

The histories of the TPMs examined in this chapter exhibit some commonalities among their significant differences. A TPM often begins in a small, entrepreneurial environment, such as small, organic farmers; small wind and solar energy developers; non-profit and entrepreneurial recycling firms; small, socially responsible investment firms; and independent retail stores that sell fair-trade products. In the case of infrastructure, the driving force was a professional reform movement, but even in that case a segment of the TPM was organized as entrepreneurial, small firms that developed the alternative designs. Generally, there is an established industry that attempts to block the innovations of the TPMs or simply ignores them as unimportant, such as the mainstream agrofood complex and retail supermarket industry, grid-based electrical power companies, governments oriented toward landfilling and a packaging industry oriented toward incineration, sprawl-oriented developers and growth coalitions, the mutual fund industry, and the mainstream retail industry. TPMs can sometimes leverage their position by obtaining support from a countervailing industry or set of firms. Examples include the linkages between small organic farmers and natural foods stores and retail food cooperatives, large energy and technology firms willing to manufacture solar panels and wind turbines, the waste management industry for recycling, and neotraditional developers for New Urbanist reform plans. In the financial field, religious and educational institutions provided enormous support in the early stages of socially responsible investing and consumption. Support from the countervailing industries may also be blended with some government support, as in the case of renewable energy portfolio standards, recycled content procurement standards, and funding for low-income housing and transit development projects that are consistent with reformist principles.

As the market develops, the established industries shift some investments toward the new opportunities, either through their own product development or through acquisitions of smaller firms that developed the alternatives. In the case of fair-trade product labels, development took place in non-profit organizations, which cannot be acquired. As the recolonization of the newly diversified market develops, object conflicts shift from the stark contrasts of the alternative technology and product versus that of the established industry to the more complex choices of a continuum of alternative, complementary, and mainstream technologies and products. The new shape of the technological field is an outcome of the market growth of the alternative products, the incorporation of the alternatives into mainstream industries and markets, and the complementarization of the design of alternative products that occurs when the mainstream industries accept but transform the alternatives. The redesign of the alternatives tends to make the alternatives complementary to existing industrial practices and markets. Examples are the transformations of organic food from fresh, local, directly sold products to organic processed food sold by large firms through supermarkets, the transformation of off-grid renewable energy into wind and solar farms or distributed energy, and the transformation of transit-oriented urban development into the neotraditional suburban development.

Just as IOMs tend to achieve partial success through the politics of the partial moratorium, so TPMs achieve partial success through the politics of the incorporation and transformation of the alternative technologies and products into countervailing and established industries. In some cases the industrialization of the alternative product has resulted in a regrouping of the movement and relabeling of the object. In other words, the movement side of the TPM may reject what it sees as the cooptation of the original vision, and it may shift increasingly toward more carefully defined projects in support of local control and low-income access. In some cases, such as the increased emphasis on localism among organic farmers, the incorporation and transformation of the TPM is closely connected with the development of a parallel localist pathway.

Throughout the history of the TPM there are ongoing object conflicts or definitional struggles, and they are especially pronounced between a social-movement vision of material culture and a for-profit, corporate vision. When large industries incorporate the object and make it compatible with existing products (often turning it into a green product for a targeted market niche), the story does not end. One of the most consistent and readily evident sites of ongoing object conflicts is the politicization of consumption through the politics of the boycott and buycott. As the technological field diversifies, the original alternative product becomes differentiated into a complex and confusing field of consumer choices, such as fresh, local organic or fair-trade organic versus processed, industrial organic or new labels constructed by corporate retailers.

Object conflicts also occur over regulatory policies, industry standards and the agendas for research fields, and even the existence of the new technologies. Regulatory standards are subject to constant pressure for redefinition, especially from industrial corporations that can benefit from a diluted standard that still generates a marketplace premium (as in the case of organic standards, renewable portfolios, and green building codes). Likewise, research agendas follow the money, and as industries articulate their own vision of the transformed alternative object, the associated research and design fields diversify to include research on the more complementary and less social-movement-oriented visions of the objects (as occurred in the solar and wind industries). In some cases the new, complementary objects and industrial processes (such as grid-based wind farms or recycling and resource recovery facilities) can produce new oppositional movements, so the redesigned form of the object generates a new round of contestations.

One strategy that a TPM can use to exert some control over the direction of object conflicts and the contours of the technological field is to establish labeling standards for the product. The history of organic standards suggests how the original, entrepreneurial segment may lose control over the standard-setting process as it proceeds to national and international levels. Likewise, a technically oriented, production-based standard makes it relatively easy for social-movement values such as local ownership or fair wages to become lost as the market develops. The more socially oriented, production-based standards of the fair-trade movement can maintain a social-movement vision, but they may also be watered down and displaced by alternative labels based on corporate constructions of production standards, as is occurring with coffee. In short, TPMs can rely on standards and labels to enforce a politics of product design and market share, but the more profound the vision of justice and sustainability that is built into the labels, the more likely it is that the mainstream industry's complementary standards and labels will crowd out the original vision.

As a political strategy for social change, advocates of the TPM strategy should recognize that they are playing an intense dance of cooptation and success, incorporation and transformation. Success without marginalization will almost always entail high degrees of transformation. Some advocates defend the strategy because it leverages changes in powerful industries dominated by large corporations, where electoral political strategies have been abject failures and overburdened opposition movements have been ineffective. To date, however, the alternative and complementary products of the TPMs have achieved only small market shares in the United States. (In Europe, where some governments have supported sustainable products, the record is more optimistic.[72]) TPMs should be recognized as important laboratories for the development of technologies that can reduce ecosystem withdrawals and deposits, and levers that can be applied to bring about a limited and contingent greening of industry. However, without strong government support TPMS are not likely to achieve the vision (articulated by activists) of a full-scale technological conversion of society. Apparently it will be a long, long time before the vast majority of food is grown organically, energy is produced through wind and solar, all waste is reused as industrial inputs, development is configured as compact and energy efficient, investment is in socially responsible products, and consumption is based on fair-trade principles. Yet without the TPMs such changes may never happen.

6

The Localization of Activism and Innovation

Students of globalization have noted that the internationalization of economic, political, and social relationships has coincided with a paradoxical relocalization of society. One aspect of localization, the development of regional industrial clusters, is particularly attractive to cities that have lost traditional manufacturing industries to foreign competition and wish to maintain a base of high-technology, well-paying jobs for their regional economy. Where there is a full infrastructure of government support, high-quality universities, service firms, and institutions that foster firm-to-firm networking, a particular city or region can compete globally as a hub of innovation in one or more industries. Although the best-known cases are financial hubs such as New York or technology clusters such as Silicon Valley, there are also a few instances of what I call a "green technopole." In other words, high-tech clusters of environmentally oriented industries, such as new energy technologies, have begun to emerge.[1]

Whereas the high-tech manufacturing model of the technopole links the local to the global through the development of clusters of export-oriented industries, an alternative approach, localism, links the local to the global through regional policies based on import substitution. Although the two strategies are not mutually contradictory, they tend to appeal to different regions and to different constituencies within regions. Whereas large cities with existing industrial clusters and substantial research infrastructures are better positioned to develop the export-oriented, high-technology cluster, cities that lack the resources to develop industrial clusters, such as the smaller cities of the American "rust belt," may be more drawn to what Michael Shuman has called "locally owned import substitution." Cities that have lost employment through outsourcing of jobs and closure of corporations have, in some cases, become skeptical of the value of chasing after another large, multi-national corporation that may decide to stay only as long as the tax breaks last. Instead, some city governments have paid attention to job growth that occurs in organizations that are more deeply connected to the region: employee-owned enterprises and cooperatives, locally owned small businesses, non-profit organizations, and public-private partnerships. Gar Alperovitz's *America beyond Capitalism* and Michael Shuman's *The Small Mart Revolution* are two examples of research that is charting and defining the emergent politics of economic localism.[2]

In addition to the high-tech manufacturing cluster and locally oriented import substitution, a third way in which globalization is linked to the relocalization of the regional economy is through changes in the welfare state in the wake of increasingly tight government budgets. After World War II the United States led the world in manufacturing, but the restoration of Europe and Japan and the growth of manufacturing in newly industrialized countries weakened the country's manufacturing base. By the end of the twentieth century, most of the economic growth in the United States was in the service, retail, financial, and related non-manufacturing industries. The erosion of the country's global economic position, the aging of the population, and increased health care costs have created substantial pressures on the welfare state arrangements of the New Deal and Great Society eras. The federal government's spending on entitlements has shifted in two major ways: devolution to state and local governments, often through block grants that tend to be reduced periodically, and privatization, or the search for public-private partnerships and NGO intermediaries that can step in where the welfare state has stepped out. The combination of devolution and privatization has tended to shift the attention of activists from the federal to local politics and from activism to service provisioning. As the politics and the economics of the welfare state have become more localized, new organizations and partnerships have emerged to bridge the gap between human needs and state provisioning.

Both the import-substitution strategy and the shifts in needs provisioning articulate a politics of locally oriented democratic control and access in a world that, at least to many community leaders, appears to be increasingly governed by large corporations, by global institutions of finance, and by national governments that are not responsive to the grassroots desire for justice. As a result, the era of globalization has created the conditions for another group of alternative pathways: those that engage in the politics and economics of localism, and those that address the problem of access for the nation's poor in the context of devolution and privatization. (See table 6.1.) In general, there is less focus on the politics of technological fields among the alternative pathways of localism and access than among IOMs and TPMs. However, the localist organizations have developed some radical innovations in the ways that technology and infrastructure are configured, and likewise the access pathways have often been incubators of organizational innovations that may be of general value in thinking through a realistic organizational basis for a society that would do a better job of addressing the pressing issues of environmental sustainability and social justice.

Localist and Access Pathways

Although one might think of localist and access organizations as legacies of 1960s countercultural experimentation and the civil rights and feminist movements, they often bridge conventional left-right political divisions. For example, the politics of localism may articulate traditionally conservative agendas of opposition to big government and

Table 6.1
Alternative pathways: localism and access.

Fields of action	Localist pathways	Access pathways
Food and agriculture	Local agricultural networks	Anti-hunger, community gardens
Energy	Public power, community choice, home power	Fuel banks, weatherization
Waste and manufacturing	Reuse and resale	Thrift
Infrastructure	Local sources, cohousing, ecovillages	Community development corporations, cooperative housing, transit access
Finance	Credit unions, community currencies, local labels	Community development credit unions, micro-enterprise finance, time banks

international governmental organizations with traditionally progressive agendas of opposition to control by distant multinational corporations and global institutions of finance and trade. Likewise, access organizations may have historically operated in the tradition of New Deal and Great Society liberalism, in that they hold government responsible for ensuring that the minimum needs of its citizens are met; however, under neoliberal political conditions access organizations sometimes endorse self-help activities, such as occurs in community gardening, and they also seek partnerships with the private sector. Many of the access pathways have substantial involvement from faith-based organizations, where progressive political demands on the state may be replaced by the frames of private-sector charities and family values. In view of the diversity of the organizational missions and frames, "alternative pathway" again seems a more appropriate term than "social movement." Nevertheless, there are frequent points of contact with social movements, as I shall argue in the historical analyses that follow.

From an organizational perspective localist institutions include community-controlled public agencies (such as community-owned electric power utilities and regional public transportation systems), locally owned businesses, alternative living arrangements, public and non-profit schools, faith-based organizations, and other non-profit organizations. Access institutions are generally NGOs, faith-based organizations, public service agencies, or public-private partnership organizations. The missions of localist organizations may range from a narrow focus on revenue and profits in the small-business sector to public service, political reform, and education, and the missions of access organizations may range from a political reform agenda that is connected with broader social movements to a charitable agenda that is connected to faith-based organizations. A publicly traded corporation that is headquartered in a region is considered here to be "local" but not localist. It may contribute enormously to the region's economy and to its non-profit sector through donations, but the owners are geographically dispersed shareholders, and the mission of the organization is restricted to maximizing returns for shareholders.[3]

Localist organizations produce earned income from the sale of goods and services, but even the for-profit, locally owned small businesses have the opportunity to emphasize organizational viability and workforce stability over short-term earnings growth. The exception is startup companies that intend to initiate public stock offerings, so the phrase "locally owned small business" will be used to refer to small businesses that intend to remain locally owned. Given their ability to deemphasize earnings and growth as the core mission, there is a potential for locally owned small businesses and other localist organizations to embrace the values of environmental sustainability, community service, and just working conditions.

Whereas localist organizations exhibit a diversity of forms and missions that is similar to the diversity of the TPMs, access organizations typically are non-profit civil-society organizations. Access organizations tend to look more like IOMs, but they are less likely to engage in protest politics. Although there are some instances of recourse to protest, especially in the early phases of some of the access organizations that will be discussed, a more salient repertoire of action is advocacy for the poor and the development of funding for service delivery. The organizational forms of the access pathways contrast with those of localist and TPM organizations, which tend to be characterized more by for-profit firms. Partnerships do occur between access organizations and private-sector firms, but the firms are generally sources of donations and not the primary organizational site of the history of an access pathway. In place of the role of for-profit firms, many of the access organizations have religious roots, and religious support and background can motivate both a social-change orientation and a less political, charitable orientation. Some of the access action can also be traced to the social movements of the 1960s, including the American civil rights movement, in which religious influences were also prominent.

From one perspective, access organizations represent the redevelopment of civil society in the wake of state retrenchment; in particular, they address the charitable dimension of civil society that has historically been associated with service provisioning by churches. From another perspective the organizations represent a continuation of poor people's movements, often at a local or state level as a result of welfare devolution. Just as the development of localist action in the last years of the twentieth century and the early years of the twenty-first draws considerable strength from local frustrations with the loss of economic control and well-paying jobs associated with global outsourcing, so the development of access organizations during the period draws strength from the retrenchment of the federal government's commitments to the poor.[4]

In addition to comparisons based on political ideology and organizational forms, localist and access action also can be compared from the perspective of their goals. In the pure form of complete import substitution, localism involves local ownership of non-franchise or non-formula businesses, local production of goods and services, local inputs into production, and sale of goods and services primarily to local markets. In practice there is a wide spectrum of organizations that approximate one or more of the features, and there is a shared sentiment in favor of rebuilding local economic sovereignty in the wake of dislocations that have occurred as a result of industrial consolidation and globalization. As large corporations have left town and left behind empty buildings and jobless families, economic localism has emerged as a banner under which communities think through ways to redevelop their economies without being held hostage to yet another multinational corporation. The leftovers of a de-industrialized economy (e.g., local civil-society organizations, small businesses, and local government agencies) can be rearticulated to form their own small-scale clusters. In the process localist advocates can draw on long-standing political culture frames of Jeffersonian self-sufficiency and the romance of owning one's own business.[5]

Although localism can draw on deeply held values in American political and economic culture, its strategy of import substitution can sometimes clash with the values of self-sufficiency. From a neoliberal or even neoclassical economic perspective, a nation-state's use of trade barriers to support industries based on import substitution is generally rejected because it is seen as assisting inefficient industries. Critics of import-substitution policies argue that protectionist industrial policies will tend to create corruption and non-competitive industries. However, in the case of cities and regions import substitution does not involve government-sanctioned trade barriers as much as consumer education about the positive multiplier effects on the local economy that occur when consumers engage in the buycott of locally owned businesses. A buycott of locally owned businesses often coincides with the boycott of the "big box" superstores, at least for purchases where such substitutions can be made easily. To the extent that the localist strategy can claim success, it tends to be found more in service industries, food and agriculture, energy, retail, and banking, but it is less evident in the manufacturing industries. In other words, in industries where production costs can be reduced significantly through economies of scale, there will be a need for higher levels of capitalization and an advantage for firms that compete in broader markets, a condition that will tend to favor the publicly traded corporation and the developmental pattern of the technopole.[6]

Access action also can come into conflict with deeply held values of self-sufficiency. In a political culture that emphasizes hard work and competition, the charitable dimension of access action can been framed as support for laziness. The countervailing religious frames of support for the less fortunate can be mobilized to counteract the "welfare chiseler" image, but access action can also be framed through a more secular diagnosis of dislocations caused by outsourcing, layoffs, abandoned factories, and other effects of globalization. When framed as a response to globalization, access action is directly analogous to the import-substitution rhetoric of localism (in which self-sufficiency discourse is shifted from the individual to the community, which is to regain sovereignty in the wake of dislocations caused by the global economy). In this way, the context of globalization can unite the older religious charity frame and the potentially negative self-sufficiency frame. However, for the connection with globalization to be credible, access action has to be framed as temporary work that is needed to overcome what the community hopes will be temporary economic dislocations. One way of overcoming those dislocations is through a localist strategy of economic development, and consequently one can sometimes find an overlap in local networks on the issues of local economic control and local poverty. The networks are especially developed in the food and agricultural field, where local agricultural and anti-hunger networks have become interwoven.

Localist and access organizations can also be compared regarding their relationship to innovation and design that is more central to IOMs and TPMs. Because localism and access are primarily about justice in the sense of retrieving control over a community from transnational economic institutions or providing basic goods and services to those in need, they do not necessarily imply a concern with environmental sustainability that is characteristic of the environmental IOMs and TPMs discussed in the two preceding chapters. In fact, the environmental politics of localist and access pathways can be quite brown or at least disconnected from environmental issues. Green localism is a possible form of localism, but it should be clear that the primary goal of localism is not necessarily environmental remediation or even a radical rethinking of environmental design. Likewise, access institutions may develop a sustainability dimension, but it is generally a secondary goal or dividend.

Just as localist and access pathways are not necessarily connected with environmental agendas, so localist and access agendas can be disconnected from each other. Localism does not necessarily imply a concern with access for the very poor; instead, the most enthusiastic support for localist politics tends to be in the small-business sector, such as Main Street retailers who have organized against "big box" superstores. Conversely, access organizations often do not address ownership issues such as local economic democracy and control, which might be seen as confrontational to corporate donors. Spaces are created within localist institutions to develop broad missions that include environmental values and poverty remediation, and poverty amelioration can be addressed through local entrepreneurship, as occurs in the microfinance organizations. In other words, converges occur, but as with other types of movements and action such connections are historically contingent.

Finally, localist and access action can be compared on the issue of incorporation and transformation, the topic that again will be the focus of the historical analysis. The incorporation and transformation process for localist institutions is to some degree continuous with the general pattern of industrial consolidation and the transition to non-local ownership that has become especially salient for the private sector as a whole in the globalization era. The pressures are particularly intense in manufacturing, where it is difficult to produce a competitive product that relies on local inputs and sells mainly to local markets. No matter what the industry, localist organizations may undergo a bottom-up process of consolidation through franchising and outright acquisition by non-local, publicly held corporations. In my experience the driving force of localist movements is generally locally owned retail businesses and local farms that are struggling against the consolidation process. However, localism can be more than a rear-guard action against the superstore, factory farm, and other incarnations of consolidation and globalization; there are other ways of integrating locally owned, independent, non-publicly-traded businesses with broader markets. For example, in Italy small manufacturers have stayed alive by uniting through producer cooperatives, which can reduce input costs through bulk purchasing and increase access to distant markets. Likewise, consumer product labels, such as the fair-trade label discussed in the previous chapter, can encode organizational structure and localist values in the product label and integrate small-scale, localist organizations into global markets.

Because access organizations are not businesses that can be acquired by larger, publicly traded firms, the incorporation and transformation process is less direct than for the locally owned small business. However, I will provide evidence that access organizations have undergone routinization from advocacy work and activism to service provisioning. The incorporation and transformation process for the access organizations tends to be closely linked to the downward shift of federal government responsibilities to the states and local governments and the outward shift to the private sector as a source of charitable donations and to NGOs as a source of service provisioning. As a result, the typical process of development in the access pathways is an increasing preponderance of charitable or service-provisioning organizations, often in partnership with private-sector support and government cost sharing. Individual organizations may also undergo mission drift from advocacy or activism to charity, but when faced with reductions in government support, they may remobilize in an advocacy mode.[7]

Food and Agriculture

In the United States the decline of the small family farm is advanced, and corresponding localist action is also very well developed. In part the consolidation process can be traced back to the failure of nineteenth-century homestead policies, which limited farm size to 160 acres, too small to sustain a family in the arid conditions of the Great Plains. The disconnect between policy and ecology set the stage for the rapid consolidation that occurred during the period of the Dust Bowl and Great Depression. However, larger farms also had higher productivity due to increased capital-to-labor ratios and greater externalization of costs to the environment, and as a result they could accumulate capital more easily than the smaller farms. During the late twentieth century commodity chains became longer, and large, transnational agribusiness firms came to occupy larger portions of many agricultural markets.[8]

In the late 1970s, small farmers faced increasing debt, higher interest rates, and falling commodity prices. They carried out rallies and tractor-cades in various state capitals and in Washington, where they demanded price stability and the end to imports that competed with domestic agriculture. Naively expecting a rapid response from Congress, the farmers at first found the doors closed. However, in the next decade the American Agriculture Movement and other farm organizations did achieve some concessions from state and federal legislatures. The farmers had hoped that changes in agricultural pricing policies would put an end to the consolidation process, but the legislative changes did little to reverse the trend. The American Agriculture Movement's approach to rebuilding agriculture did not rethink the fundamental relationship between the farmer and the consumer, which relied on long commodity chains that shifted profits to distributors, processors, and retailers. As a result their effort to save the small family farm was only partially localist in orientation. It focused on the issue of retaining local ownership of small and medium-size farms but failed to rethink non-local inputs to production and non-local markets.[9]

In contrast, local agricultural networks engage in a more complete politics of import substitution by connecting regional consumers with regional farms and locally owned food businesses. Often the strategy of import substitution extends to the production process and technologies of production, where farms engage in seed saving, multi-cropping, composting, and other technologies that reduce dependence on distant corporations that supply seeds, fertilizer, and pesticides. Local agricultural networks also engage consumers in productive activity through voluntary and paid work in farmers' markets, consumer cooperatives, and community-supported agriculture farms, and increasingly they have become networked with food banks and other food access organizations.

However, before becoming too celebratory of the food and agricultural field as the leading example of localism, as well as of the increasing synergies between localist and access organizations, it is important to remember that even here localism is not necessarily green. Because the primary concern is to replace the long commodity chains of industrial agriculture with local production for local consumption, local agricultural networks do not necessarily engage in sustainable production or even production that would meet organic standards. Doing so can be expensive for small farmers, and farmers who depend mostly on direct marketing to local consumers may opt not to gain certification. Some farmers may opt to farm in a quasi-organic manner but rely on their reputation and local networks of trust to secure a price premium. Likewise, although many community gardens farm organically, the label is largely irrelevant for gardeners who do not intend to sell much of their produce on the market. At most of the farmers' markets that I have visited across the country, the majority of small farmers were not selling organic products, and likewise some of the community gardens finessed organic techniques, particularly in the immigrant gardens on the West Coast. Although locally owned farms and especially community gardens may tend to implement more environmentally sustainable practices, the connection is not necessary. In fact, in order to stay in business, some family farms have shifted into factory farming techniques and, in the case of dairy farms, the use of growth hormones.[10]

When selling directly to local markets, small farmers who sell organic food can generally capture a price premium. The strategy can be especially valuable for small farms located near urban areas, where direct links to consumers and restaurants can create a stable market with relatively stable prices. However, location near an urban area has other drawbacks that are affecting the general historical trend of consolidation. Bear in mind that in the United States urban land area doubled from 1960 to 1990, and by 2000 it occupied 20 percent of all land area. The growth had a negative effect on the approximately 30 percent of American farms that were still located in metropolitan areas during the 1990s. Farmers located near urban areas have tended to shift toward high-quality produce for urban markets or even to redesign the farm to become a source of entertainment and education for urban residents by providing access to hayrides, corn mazes, petting zoos, sheep-shearing events, tours, and other farm experiences. By changing the definition of a farm and its relationships to consumers, there is a greater chance of rescuing metropolitan agricultural spaces from land speculation. Some successful farms have even made the transition to non-profit status.[11]

From an organizational perspective, local agricultural networks include a cluster of innovations that operate at the intersections of private-sector firms and civil-society organizations. This section will discuss five major organizational types and, where information is available, briefly look at their trends: non-profit status and community land trust arrangements, community-supported agriculture, food cooperatives, farmers' markets, local food labels, and locally oriented restaurants. The developments create consumer demand by politicizing consumption as a mixture of environmental, economic justice, and health considerations. There is an enormous literature on the topic; this section will extract from the literature a discussion of the late-twentieth-century trends of incorporation and transformation.

By seeking the protective wing of non-profit status, farmers do not have to make a living wholly from food sales. New revenue streams open up, including educational fees, grants, and donations. Another alternative is to maintain for-profit status and farmer ownership but to anchor long-term development rights to a non-profit organization. Under either the non-profit or community land trust model, the farm can formally articulate organizational goals that are less like a for-profit business and more like a civil-society membership organization. As a hybrid organization that engages in both civil-society activities and in private-sector production for markets, this type of farm provides one example of organizational innovation that is suggestive of post-corporate, localist economics. The non-profit farm is also relatively immune from the general trends toward acquisition and consolidation.[12]

Another organizational development is community-supported agriculture (CSA), which applies the subscription idea to agriculture to provide consumers with regular access to quality fresh produce. Many CSA farms practice organic or biodynamic agriculture, or they offer some features associated with more sustainable farming technologies, such as reduced use of synthetic pesticides and abstention from the use of genetically modified seeds. Some CSA farms also provide scholarships for low-income members and reduced rates in return for increased work, and some donate extra food to food banks. The first CSA farms in the United States, started in New England in 1985 and 1986, were influenced by European experiments in anthroposophy and biodynamic farming as well as the writings of J. I. Rodale and E. F. Schumacher. By the early

2000s there were as many as 1,700 CSA farms in the United States. Women played a significant role in the development of the movement, especially in initiating household subscriptions. Notwithstanding the growth rate and the organizational innovation represented by the extension of the subscription idea to agriculture, according to survey research many of the CSA farms are very small and do not generate enough income for the farming families. In other words, the CSA model is growing but remains less economically powerful than some of the other examples of direct farm-to-consumer marketing. There is some evidence for locally based consolidation of subscriptions (members receive produce from more than one farm, which aggregate their output); however, to date the consolidation is handled cooperatively among the farms rather than through private-sector firms.[13]

The membership concept also applies to food-oriented retail cooperatives. The consumer cooperative movement dates back to nineteenth-century England, and hundreds of food cooperatives were established in the United States during the Great Depression, but by the end of the twentieth century few of the cooperatives from that period survived. A second wave of food cooperatives was launched in the wake of the 1960s' social movements, and by the end of the twentieth century there were about 300 retail natural foods cooperatives in the United States. Although food cooperative sales grew at a rate that was comparable to other natural foods retailers, they were hurt by the emergence of natural foods chain stores and natural foods departments in conventional supermarkets. By 2000 the 300 retail food cooperatives had a sales volume of $700 million, whereas the two largest natural foods chains had exceeded them in sales ($3 billion) and were closing the gap in number of stores (about 220 stores in all). The food cooperatives were watching the rapid changes in the marketplace with some trepidation, and some were investing in expansion as a counter-strategy. In part to combat the rise of both natural foods chains and the growth of natural foods departments inside mainstream supermarkets, food cooperatives also began to form regional cooperative grocer associations in the early 1990s, and in 1999 they formed a national association. The change suggests one type of solution to the paradox of maintaining localist institutions in a global economy.[14]

The farmers' market is a fourth, and arguably more successful, example of localism in agriculture. The term "farmers' market" covers a wide range of management structures, all of which share direct marketing between farmers and consumers as the defining feature. Many farmers' markets include a mixture of farmers and retailers who purchase goods from farms, and in some cases farmers themselves purchase goods for resale or travel long distances to sell their goods. Certification of local farms, such as the program in the state of California, and farmers' markets rules can enforce direct marketing of food by local producers. Like the Depression-era food cooperatives, farmers' markets declined in the post-World War II period. The Farmer-to-Consumer Direct Marketing Act of 1976 allowed the Cooperative Extension Services of the federal government to help build farmers' markets, and the agency's assistance helped spur the resurgence of the institution. From about 300 farmers' markets nationally in 1970, the number grew dramatically during the 1990s, reaching about 3,700 in 2006. A study completed in 2000 found that 19,000 farmers were selling only at farmers' markets, and 82 percent of the markets had achieved financial viability. Sales at American farmers' markets in 2000 topped $2 billion in the United States—several orders of magnitude larger than CSAs, and about three times the aggregate sales of food cooperatives, but still smaller than the natural foods retail chains. Farmers' markets also serve as catalysts for growth and incubators of new businesses, and they provide a setting for various civil-society activities, such as health and education programs as well as some political organizing around food politics. There are also some instances of convergence with access goals: some markets are located in low-income neighborhoods, and some accept food stamps.[15]

Labeling or branding local food is another way to enhance local agricultural networks, but it is considerably less developed in the United States than some of the other institutions discussed here. In many regions there is some marketing and information sharing for local farm products. One fairly well developed example is in western Massachusetts, where an organization developed the Local Hero campaign, which labeled local farm products and helped the farms to connect to consumers through farmers' markets, restaurants, and grocery stores. In other places websites provide information about the value of buying local, and they link consumers with farms, farmers' markets, and CSA opportunities. To get a sense of the potential for local labeling, one must examine the county of Fürstenfeldbruck near Munich, where the Brucker Land label met its goal of 5–25 percent market share for local agricultural products such as milk and bread. Although the local label spread to eight other counties in the region, sales of local products also appear to have reached a point of market saturation. The German case suggests that the market for local agriculture may hit a ceiling at about 25 percent of the total food market for any particular commodity, barring institutional trans-formations that would be needed to bring about more profound changes in consumer and retailer preferences.[16]

Non-profit farms, subscription agriculture, food cooperatives, farmers' markets, and local labels are examples of the increasing organizational complexity and scope of local agricultural networks, but they tend to assume a model of food consumption that involves home cooking. The model is out of synch with social trends, which indicate that the percent-age of meals eaten away from home doubled in the last 20 years of the twentieth century. Increasing numbers of restaurants have begun to feature local food, and some of the more famous restaurants, such as White Dog Café in Philadelphia and Chez Panisse in Berkeley, have become nodes in local activist networks.[17]

At least two restaurant associations now certify "green" practices. In 1990 the Green Restaurant Association was founded to provide informa-tion and certification based on a definition of greening broken down into eleven categories, including "sustainable food," recycling, energy conser-vation, and the use of chlorine-free paper products. The association offered certification and a logo for restaurants that committed to some areas of the greening process. Here we quickly enter into object conflicts over the politics of defining a certifiable level as green. The Green Restaurant Association's standard has disengaged environmental issues from the conceptualization found in natural foods restaurants as well as from the concerns of economic localism. The Chefs Collaborative, an association that was founded in 1993, was much more oriented toward sustainable local agricultural networks.[18]

Although the local agricultural networks exhibit tremendous innova-tion in organizational forms and synergies among diverse organizations, small and medium-size farms may still find that they are unable to survive financially when they rely on a strategy of only local sales. Agricultural cooperatives provide one established way of linking to broader markets. Organic producer cooperatives, such as the Organic Valley Family of Farms (as well as some non-profit organizations, such as Appalachian Sustainable Development), provide suggestions of a solution to the broader problem of non-local localism: how can localism be translated into a economic strategy that is integrated into continental and international markets? The shift to broader markets is often accompanied by what is, in effect, product redesign. Farms find it more profitable to shift from raw food to food products such as yogurt or fruit spreads and to develop a label that draws attention to provenance and localism. If the farm makes such changes with its own products (that is, outside the structure of a cooperative) and demand grows significantly, it may become caught up in a growth treadmill that will eventually lead to purchase by a large food company, as occurred with Cascadian Farms. By aggregating with other small farms under a cooperative label, it may be easier for a single farm to resist the pattern of acquisition that occurs when a farm becomes, in effect, a food-processing company. Yet, as the history of the Mondragon cooperative system in Spain has shown, even producer cooperatives are not immune from change. The Mondragon history suggests that producer cooperatives that achieve large scale and compete with established industries on the global market become more hierarchically organized and adopt organizational changes that are similar to those of corporations.[19]

Turning now to access action, I will consider the two primary forms in the field: anti-hunger action and community gardening, again with a focus on trends and changes since the 1960s. The late-twentieth-century wave of charitable action and advocacy began with renewed recognition of hunger as a national problem during the 1960s. During that decade a United States Senate investigation documented hunger in the Mississippi Delta region, and Congress subsequently developed legislation aimed at ending hunger in the country. At the national level anti-hunger advocacy organizations emerged during the 1970s and the early 1980s to push for continued federal government support. Like the national environmental organizations, they diversified to occupy different niches. For example, Second Harvest specializes in food banks, whereas others emphasize political action, such as the Food Research and Action Center, RESULTS, and Bread for the World. Although there are significant differences among the organizations regarding their focus on charitable donations versus political action, during the 1990s they were able to coordinate activity and present a united front to Congress to lobby for increased food assistance for the poor. Similar work also occurred at the state level from organizations dedicated to hunger action in their region.[20]

Many of the hunger advocacy organizations have chapters across the country that take up charitable work at a local level. In the 1960s a new institution, the food bank, emerged to supplement the long-standing work of soup kitchens and other food-related charity. Anti-hunger services and advocacy grew rapidly during the early 1980s, when a confluence of severe economic recession and spending cuts on social programs created a wave of new poor. Under legislation enacted in the mid 1980s, federal food surpluses purchased under agricultural price support programs were channeled to the poor via a system of local boards governed by religious and charitable organizations. The decentralization and privatization of food provisioning meant that anti-hunger work became both increasingly localized and linked to the private sector. Furthermore, anti-hunger work became linked to localist agricultural institutions such as community gardens and local farms, a development that made it possible for food banks to offer some fresh produce and even occasionally some organic produce. However, in the beggars-can't-be-choosy world of access organizations, access to high-quality, fresh, local, organic food is generally limited.[21]

As social scientists and activists have sometimes noted, food relief tends to reproduce relations of dependency and class deference rather than mobilize the poor in a social struggle under the legitimating framework of a basic human right to food. However, many people involved in food relief do not regard themselves merely as providing charitable food relief for the poor; instead, they see their work as contributing to a movement that has a social-change agenda of ending hunger, though mostly through repertoires that do not involve direct-action protest. Likewise, anti-hunger action is also increasingly linked to the politics of developing local agricultural networks. The convergence between localist and access goals can be seen especially in the development of food policy councils at the state and local level. In the mid 1990s the Community Food Security Council was formed to coordinate activities at the national level. The concept of "food security" departs from the charitable premise of some anti-hunger organizations by emphasizing the self-reliance frame that is articulated with robust local agricultural and food provisioning systems. In contrast to viewing food relief as an emergency measure in the wake of the failure of the welfare state to provide for basic human rights, community food security advocates point to a broader set of societal changes than the direct provisioning of the hungry with food from food banks or through government entitlement programs. The broader changes include not only nutrition education and access to community gardens, but also access to food retail outlets, to the public transit needed to reach them, and to the jobs needed to pay for food.[22]

To some degree the diversification of food access politics from anti-hunger work to community food security is parallel to the transformation of the organic food movement into sustainable local agriculture. The two successor pathways come together at a variety of junctures, such as retail food cooperatives and farmers' markets located in low-income neighborhoods, but they also intersect with the older history of community gardens, which throughout the twentieth century have been linked to what today would be called the food security of the urban poor. On first analysis community gardening may look like an ideal solution to hunger because it breaks from the dependency relations of charitable food donations and because it is consistent with the import-substitution goals of localism. However, in both rich and poor countries urban agriculture has generally developed with little government support; in fact, governments and the real estate industry have often opposed the appropriation of urban land by low-income residents. Many gardens in American cities are located on land for which the cultivators have only short-term rights. The precarious land tenure has vexed urban agriculture efforts in both rich and poor countries, and it has been a significant issue in the politics of community gardening in the United States since the 1980s.[23]

Historically, community gardening in the United States has waxed and waned over the decades, with poverty amelioration being emphasized in some decades and alternative frames (education, mental health, and wartime food supply) occurring at other times. At the peak of World War II there were 20 million gardens of various sorts that yielded 42 percent of the nation's fresh vegetables. The scope of the wartime victory gardens, some of which are still in operation, gives some sense of the level of food provisioning that can be achieved by gardening. In the earlier waves governments or corporations were the prime initiators, whereas in the post-1960s wave community gardening was much more of a grassroots phenomenon. Over time, the community gardening efforts became subsumed under the broader urban government mantles of neighborhood development, urban green space development, and even recreation.[24]

By the end of the 1990s there were an estimated 2 million community gardeners and more than 6,000 community gardens in the United States. Eighty cities had community gardens programs, but the gardens were disproportionately located in large Eastern cities. The American Community Gardening Association reported an increase in membership from 250 to 900 during the 1990s and a 22 percent net increase in the estimated total number of community gardens. The growth was slowed in several cities, particularly New York, because of increasing land values and hostile urban political administrations. The numbers are very hard to estimate, and in our interviews with community garden leaders we found that even veterans with an immense local knowledge were not sure how many community gardens were located in their city.[25]

More than any of the other institutions discussed in this section, community gardens represent a very close approximation of localist and access goals, and at the same time they generally engage in organic or quasi-organic agriculture. The gardens are often located in low-income neighborhoods, and they provide a source of fresh food for the low-income gardeners. In addition, many gardens have separate plots that are dedicated to local hunger projects, or they have programs that connect the gardens to local food banks. When one looks more closely at the networks of community gardeners and local anti-hunger activists, they often overlap in terms of biographies, personal relationships, and organizational goals. Community gardening is also a site for educational work around the value of sustainable local agriculture, and increasingly community gardening organizations see schoolyards as an area of potential expansion, where children can learn about food and even eat some of it in their cafeteria.

Because there is tremendous variation of community gardening programs across the United States, any discussion of the general historical trends during the late twentieth century is vexed by local variation.

Nevertheless, one general trend has been the formalization of land tenure, by which community gardening has gone from a kind of social-movement activity of taking back the neighborhood to a city-run program. In cities that have large numbers of abandoned lots, mostly in the East and the upper Midwest, there are still many guerilla gardens (that is, plots developed on abandoned lots without necessarily having received permission) as well as gardens located on lands that have short-term leases from the city. Where land values have skyrocketed (especially in the West Coast cities), the local government has granted the gardens long-term tenure by supporting new gardens in city parks or on other public land. Where community gardens have flourished, it is often the case that one or more non-profit organizations have played a significant role in development, technical assistance, political mobilization, and land tenure. Although it is now recognized that community gardens do much more than provide food (including helping to revitalize a neighborhood, reduce crime, and strengthen informal ties), the gardens can also help increase the value of surrounding property, which in turn creates a motivation for financially pressed cities to sell off the land. In some cases, such as New York in 1999 and Portland in 2005, large public mobilizations emerged to defend community gardens against proposed selloffs or cutbacks in city programs. Defense of the gardens can lead to protest activity, such as the case of gardeners in Sacramento who chained themselves to an apricot tree to preserve their garden, and it can also lead to repressive activity, as occurred in New York when the city bulldozed gardens overnight with no public notice.[26]

To summarize, a few general patterns can be flagged regarding the history of incorporation and transformation in the food and agricultural field. First, there is a tendency for local governments to incorporate localist and access action into their general missions, such as by providing support and facilities for farmers' markets and community gardens, or by endorsing food and hunger goals in food charters. Second, there is a tendency for anti-hunger work to become increasingly localized and based on partnerships with the private sector, due largely to the policies of welfare devolution and privatization. As the change has occurred, anti-hunger work has tended to grow into and alongside the emergence of local agricultural networks. Third, even the small and especially medium-size farms that are integrated into local agricultural networks face severe financial pressures.

The financial dilemmas of small, locally oriented farms occasionally drive them to experiment with solutions to the problem of non-local localism. One approach to solving the problem is to develop a local, farm-based label and shift into food processing, but the strategy can lead to the incorporation of small food-processing operations into the food conglomerates. Another option is to connect with producer cooperatives that can facilitate entry into non-local markets, but even the producer cooperatives undergo bureaucratization, delocalization, and shifts into food processing as they increase in scale. A third solution is to transform the farm into a non-profit entity and to broaden the mission from food production to service provisioning such as education, entertainment, and anti-hunger work.

Other than the direct purchase of small, independent processors of natural foods by global food-processing firms and the development of producer cooperatives that market local or family farm products, there has been little incorporation of local agricultural networks into the mainstream food and agriculture industry. The large supermarket chains, including the natural foods chains, may occasionally include local agricultural products, but they have not yet exploited the local niche systematically. Some upscale restaurants have done so, and one restaurant owner in Vermont had discussed a plan to set up a food franchise based on local food, but so far the restaurant chains have not entered the market systematically. The category of locally grown food may not yet have developed the consumer demand that organic food now has, but the higher transaction costs, uneven supply, and lack of product uniformity may also be preventing the incorporation of local food into the mainstream food industry. The products of sustainable, local food networks could easily become just one more market niche in the supermarket, as could local food franchises in our fast food culture.[27]

Regarding the effects on the technological and industrial field, both localist and access pathways have been a source of significant organizational innovation. New institutions such as CSA farms and food banks have emerged as alternative organizational models for societal food provisioning, and community gardening has been repositioned as both an access institution and a factor in the broadening of the traditional mission of the parks and recreation departments of city governments. Some small farms and community gardens also serve as experimental laboratories for developing organic agricultural and horticultural techniques. The farms and gardens of sustainable local agricultural networks preserve the social-movement vision that was more prominent in the early phases of the organic agriculture TPM, and they also continue to develop the agricultural and horticultural knowledge and technology that is being lost as the organic agriculture industry undergoes scale expansion and dilution of production standards.

Energy

Localism in the energy field involves two main pathways. Public power refers to local government ownership of electrical power generation, transmission, and/or sale, but the discussion of public power will also include the emergent phenomenon of aggregation of municipal customers for bid to private-sector utilities. Although there are other types of energy (such as natural gas and oil for home heating), the discussion will be limited to electrical power. The second localist pathway is off-grid or home power, which has increasingly merged with the phenomenon of distributed energy.

Local government ownership can be found in other localist pathways, such as government ownership of farmers' markets and the use of public land for some community gardens. In the energy field, the local government's role includes, in some cases, the management of power generation and transmission. Public power is not the only type of municipal electricity in the United States; there are also cooperatives, owned by consumers and usually found in rural areas, and investor-owned utilities, which are regulated private corporations. Although public power and rural cooperatives constitute the vast majority of utilities in the United States, in most states the largest cities are served by investor-owned utilities. As a result a relatively small number of investor-owned utilities serve a relatively large number of customers. Furthermore, even many of the publicly owned utilities do not own generation facilities; they merely control transmission and distribution of electricity that they buy elsewhere. However, a few large cities are served by publicly owned utilities that also control significant generation and transmission resources. Those utilities can serve as a good point of comparison with investor-owned utilities.

As was the case with locally owned small farms, publicly owned utilities are not necessarily green. If they own generation capacity, they are often hostage to their own long-term investments in fossil fuels, and when they purchase power from private generation sources, the power that comes from a grid mix can be heavily weighted toward fossil fuels. However, there are certain valences about public power that tend to shift it away from fossil fuel as the energy source. Public power companies have priority in purchasing power from federal generation sources, which include nuclear power and large, hydroelectric power. Although both are highly controversial from an environmental perspective, they are not fossil-fuel sources of electrical power. Furthermore, a few of the public power organizations (notably in Seattle, Sacramento, and Austin) have established leadership in the environmental area by making investments in or purchases of renewable energy from wind farms and distributed solar. Beginning in the 1980s they provided a model of a transition toward greener sources of electricity that could serve as a yardstick for the performance of the investor-owned utilities. Seattle City Light was able to mix hydroelectric purchases with other renewable energy generation and carbon offsets to become the first carbon-neutral electric utility in the country.[28]

Public power can also provide significant benefits to a city or region through contributions of profits to government budgets, assistance in low-income energy support, and contributions to energy conservation programs. In view of the benefits, it is not surprising that in the late twentieth century several dozen cities opted to municipalize electricity, and some also fought to retain private ownership against advocates of privatization. An example is the battle to retain public ownership over Muni Light, subsequently known as Cleveland Public Power, which was the defining fight of the career of Dennis Kucinich as mayor of Cleveland. It was not until the 1990s that he was vindicated, and the city recognized the benefits of public power. In the case of San Francisco during the 2001 and 2002, there were heated electoral battles over efforts to municipalize the utility, but the public ownership advocates were heavily outspent and narrowly defeated.[29]

Another development that is analogous to public power is the shift to community aggregation of electricity purchases. In the wake of the lost battle to take back control of power from the investor-owned utility in

San Francisco, activists helped to develop state-level community aggregation legislation, which allowed the city to purchase its power from alternative suppliers and to mandate a higher level of renewable energy. Community choice aggregation has significant potential to develop into a national "local power" movement, to use the phrase of its leading advocate, Paul Fenn. When connected with bond issues, as occurred in San Francisco, community choice aggregation can result in the construction of significant new renewable energy generation capacity. As of 2006, community choice legislation had been enacted in several states.[30]

Under either localist strategy (municipal control or community choice contracts) communities are better able to shift their energy mix from fossil fuels that come from distant sources to locally generated, and sometimes locally owned, renewable energy. Unlike green pricing programs, which rely on individual decisions, public power and community choice contracts can shift allocations at a larger scale. The shift to locally generated, renewable energy can range from purchases from regional wind farms owned by publicly traded corporations to construction of municipally owned wind farms to rooftop solar and other forms of small-scale, distributed generation. In turn, distributed generation is a point of connection with the second main alternative pathway for localism in the energy field: the more individualistic movement variously described as home power, off-grid energy, and distributed renewable energy.

Among the developments of the counterculture of the 1960s and the 1970s was a "back to the land" movement of up to a million people who set up farms and homesteads in rural areas. Although they often had little experience at farming, subsequent survey work indicated that most of the "back to the landers" who stayed on the land had other sources of income, such as contracting work or income from previous savings and investments. The mix of contracting work skills and additional income allowed them to experiment with off-grid energy. Some chose remote locations where they were unable to establish connections to the energy grid, and others had access to grid connections but for political and lifestyle purposes wanted to achieve energy independence. Together with hobbyists who enjoyed tinkering and homeowners who wanted to take advantage of tax credits, a self-denominated "home-power" movement arose. The designation "home-power" today is somewhat of a misnomer, because many of the examples discussed in the magazine *Home Power* are installations for businesses, non-profits, and local government buildings. The term "off-grid power" is historically a better descriptor, but increasingly the power is grid-connected and can even be sold back to the grid for revenue as distributed energy. Nevertheless, the localist implications of the term "off-grid" are worth noting, particularly in comparison with agricultural technologies that refuse the agro-industrial "grid" of inputs such as pesticides, fertilizer, and seeds. Furthermore, as with the local agricultural networks, off-grid or home power is not necessarily green. Many of the rural homes that have gone off grid do so by combining wind, solar, and hydroelectric microturbines with propane gas, biodiesel, and wood-burning stoves.[31]

The countercultural dimension of the home-power movement can still be found at Woodstock-like solar festivals, where families camp in open fields, visit the booths of contractors, attend lectures, teach their children how to build solar panels, and listen to solar-powered music. As in farmers' markets or community rummage sales, marketplace transactions with local suppliers take place in a broader system of non-monetary exchanges. The events do more than bring together homeowners with suppliers of renewable energy equipment; they include topics such as state energy policy, citizen action, home energy conservation, and organic foods. As a result this setting of localist marketplace transactions also creates synergies across pathways and fields.[32]

There are no accurate statistics on the number of people who have converted their homes or businesses to off-grid or distributed energy, but the main magazine of the home-power movement in the United States claims to have experienced ongoing growth in readership, from 4,000 at its inception in 1986 to 180,000 by the mid 2000s. The "tinkering culture" of the home-power movement also created the conditions for a flourishing industry of contractors, who provide installation services at a local level, and manufacturers, who specialize in small-scale energy-related technologies for the niche market. In the late 1990s and the early 2000s, the prospect of having a backup source of energy became increasing attractive for a much broader range of consumers who had suffered energy disruptions and continued price increases from their grid-supplied energy sources. In the mid 1990s, British Petroleum Solar and utility companies from the western states entered the home-power market, and by 2003 one of the major home-power catalogs offered photovoltaic panels from four companies, of which two were major energy corporations (Shell and British Petroleum). Again, we bump into one of the limitations of localism: while service installation has remained in the hands of locally oriented small businesses, the manufacturing industry, even for the specialized market of small-scale photovoltaics, has not.[33]

Although the home-power movement channels its politics through the repair and remodeling of homes and small businesses, and thus is a combination of hobby and small-scale investment as much as it is a movement, it also involves some political mobilization to change regulatory policies. Local building and zoning codes have been an ongoing area of contestation, and there have been many state-level battles to win legislative support for grid sellback (that is, the right to spin the meter backward and sell excess electricity to the grid). The power companies have resisted what advocates of home power see as a right, and in some cases home-power advocates have engaged in activist tactics such as making illegal hookups known as "guerilla solar," or a form of civil disobedience conducted in the name of the right to sell energy back to the grid. In response to popular pressure, many states have legalized grid sellback, and by 2000 the battles had shifted to more subtle conflicts over the utilities' technical specifications. Advocates of home power argued that the excessively high technical standards for grid connection constituted a de facto blockage of grid sellback rights, which legislative and regulatory changes had legitimated. Certification of installers and of distributed electricity systems appears to be a solution that satisfies the grid's need for consistency and the homeowner's right to sale.[34]

Conversion of even a fraction of the energy supply of a home or a small business to renewable energy generally represents a significant investment of time and money, and it should be no surprise that the social address of the home-power movement is, like that of local agricultural networks such as CSAs and farmers' markets, largely middle class. My impression, based on attending expos and reading the magazine articles, is also that the movement is largely men, just as women are more prominent in the localist projects of the agriculture and food. The difference is not hard to explain, given the traditional division of domestic labor between home repair and food preparation. Although the sustainability politics of the home-power movement are broad in the sense that the movement links localist, environmental, and lifestyle changes, the rights discourse tends to be individualistic, that is, centered on general rights of middle-class citizens or small businesses to build home power systems and sell their excess power back to the grid. There are some pro bono projects to build home or business power systems for the poor and working class, but the general direction has been projects for the homes and businesses of the middle class.

In the case of access pathways in the energy field, there is a much less well developed field of access action in comparison with anti-hunger action and community gardening. Support for people who cannot meet their home energy needs comes in part from the federal government's Low Income Home Energy Assistance Program (LIHEAP), a block grant program that was inaugurated in 1981, and coordination of local programs occurs through the National Fuel Funds Network. In contrast with some of the federal government's other assistance programs, in the 20 years since the inauguration of the LIHEAP program funding has remained relatively stable at a level between $1 billion and $2 billion. However, fuel costs and population continued to climb; as with the food banks there is not enough energy assistance available to meet the demand. Consistent with the general pattern in welfare programs, LIHEAP emphasizes devolution to the state and local level with partnerships between the private and public sectors. One result of block grants has been the development of fuel banks or fuel funds, that is, partnerships between utilities and non-profit organizations that provide low-income households with energy assistance and in some cases weatherization services. Fuel funds put together federal grants with donations from utility companies and other non-governmental sources. Where people are served by public power utilities, the utility sometimes provides direct temporary assistance to low-income households, but investor-owned utilities tend to refer customers to fuel funds and other NGOs.[35]

As in other cases the funding structure channels the work of non-profit organizations into service provisioning at the local level. However, some organizations engage in advocacy work at the national level, in ways similar to some of the national anti-hunger organizations. For example, the National Low Income Energy Consortium sponsors an annual conference, and the National Fuel Funds Network mobilizes support for

LIHEAP in a Washington Action Day. The network also urges governors to develop charitable assistance programs and attempts to gain a place for the poor in policy making that involves energy restructuring. Still, on the whole the role of advocacy work is much less evident for energy access than in other fields. The histories and ethnographies are not yet written for the topic, but the available information suggests that national advocacy work has developed at least partially from the charitable work at the local level.[36]

As with food relief efforts, low-income energy assistance is generally dissociated from environmental values. Although the donated energy can include renewable sources, the goal is to provide home energy assistance and not to worry about whether the energy is green or brown. However, beyond the cluster of access work of fuel funds and LIHEAP, there are also some instances of linkages between energy access issues and environmental goals that are parallel with the case of community gardening in the agricultural field. One example is the work of the National Center for Appropriate Technology, which was founded in the 1970s and combined anti-poverty concerns with energy conservation. The resulting weatherization programs were originally supported mostly by the federal government, but over the years the utility companies have also offered low-income weatherization programs. Because energy costs are about 25 percent of household expenses in low-income families (in contrast with about 4 percent for middle-class families), low-income weatherization for the poor and working class can have a significant effect on the household budget. Furthermore, energy conservation remains the most affordable and most simply localist option in the sense of pure import substitution. Another example of the convergence of access and localist values is the emergence of organizations dedicated to green affordable housing, which will be discussed in more detail in the infrastructure section of this chapter. Most of the programs to date involve construction or remodeling that makes the home more energy efficient, but there are a few examples of low-income housing that utilize renewable energy, such as rooftop solar energy.[37]

In summary, public power and home power represent two localist alternatives to grid-based, investor-owned utility power, and community choice aggregation represents a strategy to enhance local control over investor-owned utility pricing and energy mixes. Community choice legislation represents an incorporation and transformation of municipalization as a political strategy into a market-based approach that is more in tune with neoliberalism and energy market deregulation at the federal level. However, community choice can also help cities avoid an unfortunate downside of municipalization of electric power: the high costs of purchasing transmission and generation capacity as well as developing the technical expertise needed to manage it successfully. If combined with bond issues that fund the construction of distributed generation such as those in San Francisco, community choice could eventually lead to much higher levels of decentralized energy ownership than the grass-roots home-power movement has been able to mobilize. In this sense, community choice works as a form of localist jiu-jitsu in a market-oriented regulatory environment.

Although there appears to be no trend of a more direct form of incorporation among public power agencies (that is, privatization and sale to investor-owned utilities), there is some concern that federal regulatory changes could require them to divest their power generation from transmission operations. Because the vast majority of public power agencies are not in the generation business, only a small number of the larger public power agencies are threatened by the prospect, and it would be politically feasible for the electric power industry to drive a wedge into the political alliances between the large and small public power agencies. By separating generation from distribution, the regulatory change would likely privatize public power generation and conceivably also reduce some of the more environmentally oriented innovations that the public power agencies have supported.[38]

As seen in the access pathways of the food and agricultural field, the pattern of welfare state devolution and block grants continues to be evident in the energy field, with the result that public-private partnerships are localized. Access-oriented action tends to be focused on public-private partnership programs that provide emergency funding for home heating, assist in weatherization, or provide green affordable housing. Compared with the anti-hunger organizations and the thrift segment of the resale industry, faith-based organizations are less prominent in the access pathways of the energy field. There are some interesting trends that are beginning to link access and localist action, but other than weatherization programs and a few green affordable housing projects the bulk of low-income energy assistance does not simultaneously address environmental concerns.

Regarding the influence of the alternative pathways on the technological field, public power, community choice, and home power all tend to shift the technological field, as well as the design of renewable energy options within it, from grid-supplied toward distributed energy. There is considerable potential for technological innovation for the small-scale renewable energy technologies associated with distributed generation, and there are some concrete examples, such as the biogas digesters being developed by the public power agency in Sacramento and the many small-scale projects that are documented in magazines such as *Home Power*. Likewise, concern with energy efficiency and weatherization in both localist and access organizations supports innovation in energy conservation. However, with the exception of the National Renewable Energy Laboratory, the innovations tend to emerge from the grassroots networks of users and installers. The situation is parallel with the informal knowledge and horticultural technology that is developed among community gardening and sustainable local agricultural networks, where the knowledge development takes place largely within the tinkering networks of user-producers (gardeners, farmers, installers of home-power projects) rather than in university-based or firm-based research centers.

It would not be quite accurate to end with the conclusion that localist energy pathways uniformly involve a shift in the technological field toward small-scale, renewable energy. The large public power agencies have the financial resources to develop innovations in energy technology, and they have shown leadership in areas such as distributed solar, biogas, and conservation. However, even the Sacramento and Austin agencies have opted, at least for the near term, to build natural gas generation plants to serve current and projected energy needs, and they have not yet been able to solve problems of intermittency and transmission congestion for large-scale wind energy. Likewise, as I have noted, many of the home-power projects involve off-grid energy consumption that utilizes fossil fuels and/or generates greenhouse gases. In other words, even in the model cases the shift to renewable energy and conservation appears to be a long-term process.

Waste and Manufacturing

Because of economies of scale in manufacturing, the goal of economic localism (in its ideal form local production with local inputs for local use) does not work well with recycling and remanufacturing, and consequently the treatment of localist pathways in this field will be brief. One might argue that the eco-industrial park model would be a good starting point for an analysis of localism in the manufacturing field, but eco-industrial projects in the United States to date have tended to be oriented toward agricultural technologies, such as composting, or are still in the incipient form. Efforts to establish eco-industrial parks in the United States suggest that the Danish Kalunborg model (which was built around wastes from a fossil-fuel energy plant) may not be very portable, due in part to federal regulations on solid-waste disposal and liability. Even in cities and regions that are committed to environmentally oriented regional development, projects to establish eco-industrial parks have stalled. There are some experiments with the somewhat more flexible model of regional eco-industrial networks, but they are not yet well developed. As I attempted to find examples of eco-industrial park projects in the United States, I found that the ones that had not stalled were evolving toward environmentally oriented science parks, where entrepreneurial firms are incubated with support from various levels of government.[39]

To the extent that eco-industrial parks are configured around manufacturing, they are likely to be producing for markets outside the region, and in this sense their localist orientation (import substitution of some inputs) is combined with an export orientation of products that more closely approximates the green technopole model of regional development. To the extent that demand for a firm's remanufactured product grows, the demand will tend to outgrow the constraints of local waste inputs, and the firm will import inputs from outside the region. In any case the pressure of price competition due to economies of scale will drive the firm to grow in order to compete, and it will have to seek outside capital in order to grow. Although there can be localist models of eco-industrial parks, I suspect that as they evolve they will tend to be mainly small-scale, agricultural operations such as composting for local markets.

Although eco-industrial parks are still in their incipient stage, the same cannot be said for the resale industry. The category embraces a wide variety of organizations and will be the focus of this section. Although the subcategories that I will use are somewhat fluid, in this section the growth of yard and rummage sales and the emergence of reuse centers will be considered as localist pathways, whereas the thrift industry and resale shop industry will be considered as access pathways.

The informal resale market of tag, yard, and garage sales, rummage sales, and flea markets is relatively understudied, but available estimates suggest that it is quite extensive. In 2000, 43 percent of Americans shopped at a rummage or yard sale in the preceding year, and 19 percent held a yard sale. There are no good estimates of the size of the market, but a social scientist who studies them suggested that the aggregate sales are $4 billion per year, that is, a figure that is probably larger than the combined sales of retail food cooperatives, farmers' markets, and CSAs. There is anecdotal evidence that the market is growing, and explanations for the apparent growth would probably include increased environmental consciousness, higher levels of income inequality, and perhaps even the development of home businesses through resale in collectable markets and in electronic markets such as eBay.[40]

The informal market of yard and rummage sales has developed largely apart from community development or ecological concerns. Urban governments have generally viewed the phenomenon with some suspicion, and in some cases they have instituted a permitting process to control the number of people who are running tax-free businesses out of their homes in the guise of weekly garage sales. However, there are also some cases of the incorporation of yard sales into urban policies for community development. For example, several small towns in the region of New York in which I live have used annual yard sales to attract people and to build on revitalization efforts. As an annual event, the town yard sale becomes another institution of local economic exchange that is similar to the farmers' market or solar energy festival. The town yard sale can also be framed as ecological; for example, one California city has sponsored yard sales as a way to increase the diversion rate from landfills. The examples suggest how the informal economy can be aligned with urban development and environmental goals, but the sparse literature on the topic does not provide a basis at present to judge the extent of the development or how it is changing.[41]

Another localist pathway, and one that is more closely connected with environmental goals, is the reuse center. In essence a used home supply store, reuse centers generally sell appliances, doors, flooring, lumber, windows, and other supplies used in refurbishing and furnishing. The reuse industry has a national umbrella organization, the Reuse Development Organization, which integrates a wide range of organizations with very different goals, including both for-profit businesses and non-profit organizations. Convergence of environmental and access goals is evident in some of the reuse stores that process furniture and building materials, and this portion of the resale industry most closely approximates the other environmentally oriented forms of localism, such as sustainable local agriculture and distributed/off-grid renewable energy. One example is the Chicago Reuse Center, which grew out of the first wave of non-profit recycling organizations and reconstituted as a reuse center when the industrialization of recycling displaced its position as a recycling operation. Another example is Construction Junction, which was launched in Pittsburgh, Pennsylvania, in 1999, with help from an environmental organization. Some of the reuse centers have a dual mission of diverting materials from landfills and providing housing support to low-income residents. Construction material takes up a significant portion (estimated at 20 percent or more) of the solid-waste stream that goes into landfills in this country, and many reuse centers claim to divert tons of material from landfills every week. Additional environmental benefits emerge when reuse centers offer deconstruction services. Unlike demolition, deconstruction involves the hand dismantling of buildings so that materials can be resold rather than sent to a landfill. Depending on local landfill tipping fee structures, deconstruction can be cost competitive with demolition, particularly if owners are in a financial position to take a tax write-off on the donation of materials to a non-profit reuse center.[42]

In addition to addressing environmental goals, reuse centers can become launching grounds for general neighborhood revitalization projects. Probably the largest non-profit reuse center in the United States is the Rebuilding Center of Portland, Oregon. By 2002, after only 4 years of operation, the non-profit organization grossed more than $2 million, employed 36 workers at a living wage in low-tech building deconstruction, furniture restoration, and work in the center, and had diverted thousands of tons of materials from landfills. The reuse center also helped sponsor Our United Villages, a community organization dedicated to asset-based community development and community dialogue. Another example is the Green Institute of Minneapolis, a non-profit organization that encodes localism by requiring that a majority of its board members live or work in the neighborhood. Not only has the institute established a retail store for salvaged building materials and a service company for building deconstruction, but it has also built the Phillips Eco-Enterprise Center, which includes some environmental service firms along with non-profit clients in a small-scale, green localist variant of the industrial cluster concept. Such examples of organizations in the reuse industry provide a well-rounded model of sustainability, that is, a vision that is environmentally oriented, economically viable, and concerned with access issues and community development.[43]

Neither the yard sale market nor the reuse industry shows much evidence of incorporation by large for-profit corporations. There may be a trend toward delocalization as entrepreneurs purchase household goods at yard sales (or even at reuse centers) for resale through Internet-based auction sites, but the extent of this practice has not been studied. The reuse industry is composed of a mix of non-profit and for-profit organizations, and the for-profit organizations could easily be acquired by a corporate retailer and transformed into a chain of reuse stores. However, non-profit organizations have an advantage because donated items can be written off on tax returns, and they can also serve local governments' goals of job training and development. Consequently, it is likely that the non-profit organizations will continue to play a prominent role in the industry. To the extent that non-profit organizations remain important, environmental and community-development goals will be easier to maintain.

The mission of a non-profit reuse center often includes goals of community development, job training, and access to affordable home materials, and in this sense such centers might also be considered access organizations. Even the reuse centers that have a primary environmental mission offer used goods at very low prices, and many of the stores are located in low-income neighborhoods. As a result reuse centers with environmental or even profitability goals can have a substantial access dividend, just as the reuse centers with access goals can have an environmental dividend.

In contrast, the thrift industry is driven more explicitly by the goal of access to household goods such as clothing at a low price or for free. Although in some cases environmental goals may be articulated, the thrift industry has a different emphasis even as it forms a continuum with the organizations discussed up to this point. For more than 100 years, charitable organizations such as the Salvation Army and Goodwill Industries have operated stores that serve as the equivalent of food banks or energy banks for the poor. The stores provide clothing and other household goods, including some furniture, for the poor at affordable prices, and they use their proceeds to fund adult drug rehabilitation centers worldwide.[44]

As has occurred with other access institutions discussed in this chapter, the thrift/charity industry grew rapidly since the early 1980s, due in part to cutbacks in social spending that have placed higher demands on charitable organizations. As the thrift industry grew, there is some evidence of organizational change. Information is still rather sparse for the United States, but in the United Kingdom the changes included rapid growth of charity shop chains run by organizations such as Oxfam, the Red Cross, and various disease-related foundations. With the growth of chains there was an increase in paid, professional store managers as well as formal warehousing systems. Although the thrift industry in the United States may not have reached the saturation point that appears to have occurred in the British charity shop industry, in the United States and Canada there have been similar developments of chains. In addition to the non-profit stores run by the Salvation Army and Goodwill, for-profit chains have emerged to purchase donated goods from non-profit organizations and resell them at a profit. The largest chain, Value Village/Savers, had reached $300 million in sales through 190 stores by 2000, and it was pursuing a program of aggressive expansion.[45]

In addition to the national non-profit stores and the for-profit chains, there are thousands of independent resale shops that provide second-hand clothing and other household goods at low prices. Because they are locally owned small businesses that divert consumer purchases from new stores and used clothing from landfills, they could be included in the localist economy of import-substituting businesses that have an environmental dividend. However, resale shops are also access institutions in the sense that they provide clothing, toys, and other consumer goods at very low prices and generally for a low-income customer base. The National Association of Resale and Thrift Shops, which represents about 1,000 of the 15,000 resale shops in the United States, notes that the industry is one of the fastest growing segments within the broader retail industry. Although clothing stores are a primary area of resale, furniture, music, sporting goods, and electronics are also growth areas. The organization also notes that resale is particularly healthy during economic downturns. From that observation one could hypothesize that the general growth of income inequality within the United States, which economists have associated with globalization, is a driver of the growth of the resale industry. However, in some instances used goods have also become fashionable. The trend is especially true of the used clothing stores that have changed to look more like boutiques that sell new clothing and of those that have diversified into retro clothing and other niches. Resale shops that offer high-quality, higher-priced clothing at discounts of up to 75 percent are increasingly attracting more affluent shoppers, and there is also some evidence of consolidation in the industry in the form of resale franchises.[46]

To summarize, where the localist and access organizations of the resale industry are set up as non-profit organizations, they are protected from the trend toward consolidation and incorporation into the corporate retail industry. In contrast, where they are set up as for-profit stores, there is some evidence for the growth of for-profit resale chains and resale franchises in the thrift and resale shop industries. There is a hint of incorporation of the resale industry into local government missions, albeit much less than in the case of farmers' markets and community gardens, in the few examples I found of local government use of town yard sales for economic development or environmental goals. There is tremendous economic development potential to concentrate resale stores into a district of stores with shopping opportunities that might even beat Wal-Mart on price and diversity of goods, not to mention quality. However, I have not been able to find any examples other than flea markets and antiques districts.[47]

Technological innovation is not as prominent in this field as in the food and agriculture and energy fields. The difference is attributable to the focus on the resale of used goods. To the extent that one finds technological innovation, it is more in the remanufacturing side of the field (such as the furniture remanufacturing operations in the reuse centers) or in the building deconstruction operations of the reuse centers. The latter have had to develop techniques appropriate for human-scale dismantling of buildings, and in some cases they have also had to rethink practices for equipment that is generally used for demolition. For example, rather then use a crane with a wrecking ball, as occurs in demolition operations, The ReUse People use a crane to remove the roof from buildings as a whole unit, then they lower the roof to the ground for dismantling. The shift in location of the roof reduces workplace risk and makes the task much easier for workers.[48]

Whereas technological innovation is less prominent in this field than in the small, locally oriented organic farms or the tinkering culture of the home power installations, organizational innovation is substantial. The reuse centers often combine the attributes of a small business with a mission oriented toward job training, low-price goods for low-income neighborhoods, environmental remediation, and neighborhood development. Furthermore, where reuse and resale organizations accept goods on trade or consignment, they are able to build inventories without having to rely on bank loans or angel investors. The self-capitalization strategy—which is an under-analyzed and under-exploited dimension of localist institutions—has parallels with the subscription idea in CSAs, bond issues in community choice distributed energy, and some of the community currency projects to be discussed below.

Infrastructure

The various infrastructure reform movements discussed in the previous chapter are local in orientation, but they are not localist in the sense of utilizing import substitution as a principle of regional economic development. For example, publicly owned mass transit, like the public power agency, is an example of local collective ownership for local use. However, in most cases the source of manufactured buses and trains, not to mention the source of their fuel, is far outside the region; consequently, one could argue that from a localist perspective public transit is not especially different from automobile transit. Of course, there are other local benefits of public transit, such as reduced air pollution, traffic congestion, and fuel consumption. One could argue that increased use of public transit substitutes the negative externalities of dirty air and traffic congestion with the positive externalities of clean air and clear streets, and it reduces the outflow of regional revenues for fuel consumption, and in this sense it is localist.

There is a more direct way in which the model of import substitution has been applied to public transit systems. A good example is the decision by the Chattanooga Area Regional Transit Agency to develop a local electric bus industry by giving a local startup manufacturer a contract for small, downtown circulator buses instead of purchasing buses from a distant manufacturer. Because the electric power in the city is largely from regional hydroelectric facilities, a second-order import substitution occurred through the use of electric power, which is regionally produced and is also less expensive per mile than diesel fuel. A similar case of import substitution occurs when bus systems and urban vehicle fleets select mixes of regionally produced biodiesel and ethanol for the fuels. Producer states such as Minnesota and Iowa have recognized the advantage of import substitution for the farm industry by mandating biofuel standards for their states, and the state of California has mandated that an increasing proportion of biofuel consumption must be produced within the state.[49]

Although attractive as models of import substitution, both the locally oriented bus manufacturers and the biofuels refiners will tend to expand into national and global markets in order to keep prices competitive. Consolidation is already advanced among the Midwestern ethanol refineries, which have undergone a shift from farmer ownership to ownership by large corporations such as Archer Daniels Midland. Likewise, Chattanooga's electric bus manufacturer was successful for some time, but it failed in 2004 after expanding and diversifying too rapidly. Even successful biofuel refining and bus manufacturing companies must acquire capital in order to expand, and to do so they are likely to seek non-local sources of venture capital or to become publicly traded. As this form of import substitution develops, it is likely to evolve toward the export-oriented technopole model of a cluster of publicly traded companies that are headquartered in a city but produce largely for non-local markets.[50]

A more thoroughly localist approach to transportation is voiced by advocates of walking and of bicycling. From the viewpoint of energy expenditure, human-powered transportation is much more efficient than machine-powered transportation. European cities such as Venice and Groningen provide models of the car-free city, but in the United States changes in infrastructure oriented toward human-powered transportation have been largely limited to pedestrian shopping zones, sidewalks, and crosswalks. Likewise, the intensive use of bicycle lanes found in some European countries, with their carefully designed inter-modal connections to trains and sidewalks, is only approximated in a few rare cases, such as Davis, California. Given the present confluence of low-density development, unavailability of sidewalks and bicycle lanes, presence of ice and snow in many regions, often hilly terrain, and poor physical condition of Americans, it seems likely that it will be a long time, if ever, that human-powered transportation will become more than a minor and complementary aspect of urban transportation systems in the United States. Yet it offers tremendous possibilities for a localist approach to transportation design that would also contribute to healthy neighborhoods and bodies.[51]

In the field of building design, import substitution can be found in energy conservation measures and the use of locally manufactured or harvested building materials. There are some remnants of older, localist building materials industries, such as brick manufacturing, and there are some experiments in the use of alternative, local materials, such as the use of straw bale in the Midwest and West. In addition, the Leadership in Energy and Environmental Design (LEED) guidelines offer points for the use of materials that are manufactured within 500 miles of the construction site, so there is some recognition of localism in the emerging green building standards. However, the substitution of local materials runs counter to trends toward the prefabrication of building components. Furthermore, it is likely to be much harder to convince consumers of the value of buying new homes that are built from locally made materials than it is to convince them of the value of buying local food, having distributed energy generation or community choice contracts, and patronizing locally owned businesses and credit unions. I suspect that import substitution in the construction industry is most likely to grow more through the development of the reuse center. Huge, diversified reuse stores that are the size of a "big box" hardware and home supply store are still relatively rare in the United States, but as they become more common, they will provide an increasingly attractive opportunity for import substitution, especially for remodeling projects.[52]

An alternative form of localism in housing design is focused less on the import substitution of materials than on redefinition of the domestic space to make it more consistent with a society oriented toward greater local ownership. One common criticism of New Urbanism and neotraditional housing developments is that they leave intact the fundamental housing unit of the single-family home, even as the nuclear family has given way to single-parent households and households with multiple non-parent adults. Although some of the more urban New Urbanist projects do include a variety of housing forms, the reforms in urban housing design and transportation tend to stop at the doorstep of collective ownership of housing, let alone designs that encourage domestic non-kin support networks. Cohousing and ecovillage advocates argue that residential architecture may require a much more radical rethinking of the design of the domestic space. Their projects often include sustainable design features and attempts to remove or distance the domestic unit from the automobile. Likewise, the emphasis on collective spaces helps to afford opportunities for import substitution of service purchases such as fast food preparation and child care.

To some extent the inspiration for the cohousing movement, as in the case of other post-1960s countercultural and "back to the land" movements, involved a return to previous forms of social organization and material culture that were lost during the twentieth century. Nineteenth-century agrarian villages had dense networks of extended kin groups, farm hands, and friends who lived in close proximity to each other. Cohousing represents an attempt to regain some of the lost density of social support and to provide a less stressful environment for children and parents alike. Cohousing began in Denmark in the 1960s, then spread across Northern Europe and on to North America. The movement character of cohousing is more evident in the original Danish experiments and less so in some of the American projects, where cohousing has tended to become professionalized by architects, who sometimes initiate projects.[53]

When cohousing arrived in the United States, the European movement connected with an American history of experimentation in alternative housing arrangements, but it also became caught up in the unique legal barriers of the country. In the United States cohousing groups found it relatively difficult to obtain bank financing (even from the National Cooperative Bank), especially for cooperative housing arrangements in which the entire property is owned by a non-profit corporation and individual residents own shares. Consequently, in the United States cohousing projects often use the condominium model, in which individuals own their own residences. Notwithstanding the difficulties, the number of cohousing projects grew from 27 to 64 in the period 1998–2003. Although primarily a middle-class movement restricted to people who have access to financing, there have been efforts to include units for low-income residents and to build cohousing projects for low-income groups. Likewise, although cohousing facilities do not necessarily utilize off-grid or renewable energy, some projects experiment with renewable energy and other aspects of green building design.[54]

The ecovillage movement, which emerged in Denmark, had a more explicitly political agenda and included a broader range of organizations. The ecovillage movement grew out of the Nordic Alternative Campaign of the 1980s, which linked grassroots groups and scientists through environmental and equity issues and which led to the Gaia Trust. In 1991 the founders invited representatives of intentional communities across the world to help develop the ecovillage concept, and Danish cohousing communities contributed to the movement in its incipient form. By the Findhorn meeting in 1995 there was enough organizational momentum that movement leaders decided to found the Global Ecovillage Network and to develop a self-assessment tool, in part to prevent the ecovillage concept from becoming diluted. The tool that eventually emerged provides standards along three dimensions (physical, social, and spiritual), and it serves as an interesting point of comparison for understanding the implicit assumptions of a LEED-based definition of sustainability and building design. By 1999 the movement had grown to include 160 intentional communities and 10,000 traditional villages, and by 2004 there were about 60 ecovillages in the United States.[55]

The global ecovillage movement includes some of the more political cohousing projects, but it is a distinct movement with goals explicitly oriented toward sustainability, equity, and spirituality. Ecovillage movement leaders such as Ross Jackson also articulated an explicit anti-globalization analysis, and the Global Ecovillage Network has participated in the World Social Forum meetings. Unlike cohousing arrangements but more like traditional communes, ecovillages strive to achieve the integration of work and domestic life in the same space, an ideal that is easier to implement in rural villages than in urban ones. As a result of combining work and life, ecovillages can be larger than cohousing units, and some ecovillages have reached the size of 500 people. There are some very powerful ideas in the cohousing and ecovillage movements, and some of the facilities have been constructed in urban areas with access for low-income residents, but because New Urbanist principles have won the ear of developers and government officials, it is not yet clear what would happen to cohousing and ecovillages if their ideas were to be widely incorporated into urban development and planning policies.[56]

Access action, both in the transit field and in the housing field, is permeated by the history of racial discrimination. For access to public transportation the long history of transit discrimination, which dates back at least to Rosa Parks and the early civil rights movement, has continued into the twenty-first century in the form of activism in favor of transit affordability, quality of service, and location of transit lines. As jobs and retail have migrated to the suburbs, public transportation has failed to keep pace with the changes. A basic set of statistics starkly demonstrates the extent of the problem: whereas only about 10 percent of the American population gets to work via public transit, walking, or bicycling, about 27 percent of people of color in cities use forms of transportation other than the automobile. However, as jobs follow highways to the suburbs, commuting can become difficult and dangerous for users of public transportation. The planning and placement of public transportation has tended to make it difficult for low-income residents in city centers to reach destinations, and litigation in various cities has drawn attention to new forms of transit racism that have echoes of the segregationist politics of the past. On this issue the anti-sprawl message of the smart growth movement and the transit-oriented development principle of New Urbanism have intersected with the transit racism message of the environmental justice and civil rights activists.[57]

The politics of design in the public transportation field go beyond the decision to invest limited resources into highways for the middle class versus public transportation for the poor and middle class; they also include choices among types of public transportation. Some of the urban transit agencies heavily subsidize commuter rail systems and provide a disproportionately lower level of funding for the much less expensive and more heavily used bus systems. The legacy of Jim Crow laws continues into the twenty-first century; in some cities, such as Los Angeles, the front of the bus became the new commuter rail system, and the back of the bus became the often dirty, unreliable, and dangerous public bus system. Surface light rail systems and rapid bus transit can provide an attractive compromise for yet another generation of divisive urban transit politics, and the development of such systems provides one counterexample to the argument that the politics of access are divorced from the politics of technological innovation. Rapid transit systems are attractive because of their potential to lure new riders and reduce sprawl, but they can be also designed as commuter systems that mostly benefit middle-class suburbanites and siphon funds away from the bus systems of the central city. The challenge, then, is to develop the new forms of public transit in ways that cross class and ethnic divisions and help reverse the legacy of spatial and demographic fragmentation that is a result of the highway system and of suburbanization.[58]

In the case of bus-oriented transit activism the focus on access versus environmental goals varies from city to city. In Los Angeles the primary focus has been bus access with respect to the rail system, but the Bus Riders Union also supported natural gas as a clean fuel. In Boston the early focus of an environmental justice group was on diesel bus emissions, but the group also sought to keep fares affordable. From the perspective of transit agencies the environmental and access issues can present a zero-sum choice, given the limited funds available and the significantly higher initial costs of clean-fuel buses. The solution can be diversion of funds from commuter rail projects to the bus system, but such solutions can run into opposition from middle-class transit users.[59]

Regarding access to housing, there is also a long history that can be traced back to the programs of the Progressive and New Deal eras, but for the period of interest to this study the more immediate antecedents were the War on Poverty programs of the mid 1960s. During the decade pressure from the civil rights movement and urban mayors, who built a coalition with construction-oriented unions and developers, led to the creation of the Department of Housing and Urban Development. In

1966, after walking through the Bedford-Stuyvesant section of Brooklyn, Senator Robert Kennedy worked to found the Bedford Stuyvesant Restoration Corporation. With funds from the "War on Poverty," the first generation of community-development corporations (CDCs) emerged during the late 1960s.[60]

Unlike the model of direct governmental service provisioning through public housing, the CDC was intended to involve a local level partnership between community leaders and the private sector, and it encouraged private home ownership rather than collective ownership experimentation as found in cohousing and ecovillages. Under the new federalism of President Nixon, the War on Poverty programs were transformed into Community Development Block Grants, and the focus of housing funding shifted toward rent subsidies. The Community Development Block Grant program channeled money to the mayors and was intended to open up funding to the non-poor. However, CDCs were able to recapture much of the funding, and they continued to grow from about 100 at the end of the 1960s to about 1,000 at the end of the 1970s. Narrower in focus, the bulk of the work of the second generation of CDCs was oriented toward housing and restoration rather than a mix of housing and local business development. During the 1970s general public support for federal housing funding softened as a result of financial pressure on the middle class and white flight to the suburbs, and subsequently President Reagan cut the federal housing budget from $30 billion to $8 billion. In the wake of the cutbacks CDCs sought funding from local sources and became more professionalized. Yet their numbers continued to grow, and by 1990 there were about 2,000.[61]

On the narrow metric of housing CDCs could claim substantial achievement: by the late 1990s there were more than 3,600 organizations that had built more than 500,000 homes and provided more than 247,000 jobs. However, in most cases the achievements of CDCs have not affected the general trend for urban poverty to become increasingly concentrated geographically within cities, and the CDCs that worked in the poorest neighborhoods have been fighting a losing battle against depopulation and immiseration. Furthermore, CDCs provided housing and jobs with relatively little perturbation of the system of urban class relations. Control rested not with poor neighborhoods, but with the funders, who generally did not live in the neighborhoods that were the intended beneficiaries of the CDCs. Increasingly, funding came from private sources, including large corporations. Given the finances of CDCs, they tended not to be connected with the grassroots activism of poor people's movements, and indeed they channeled anti-poverty activism into the non-profit activities of service provisioning.[62]

Because of the devolutionary policies of the 1970s, the transformation toward service provisioning occurred relatively early in this pathway in contrast with access-oriented advocacy organizations of some of the other fields. The block grant programs also had the effect of shifting the attention of the advocacy organizations to the state and local level. Some of the organizations that had focused on housing advocacy in the 1960s and the 1970s shifted toward providing housing services. As the funding declined and competition among organizations intensified, professional-ization gave some CDCs better access to funding sources than community-based organizations. The result was compromise on basic access issues such as mixed-income policies and rules that would maintain prices when owners sold and left the development. However, by the late 1990s some of the organizations showed a renewed interest in general community participation and a wider range of development goals.[63]

Another form of affordable housing, cooperative housing, also underwent changes in the wake of the spending cuts and devolution under President Reagan and subsequent administrations. Cooperative housing in the United States dates back to the late nineteenth century, and during the early twentieth century labor unions and immigrants developed affordable housing cooperatives in New York and other cities. During the New Deal and Great Society eras, cooperative housing grew with support from the federal government, but after the cutbacks of the 1980s there was a severe decline in support. Nevertheless, affordable cooperative housing continued to grow due to conversions of public housing and to the activism of tenants' rights groups and community organizers, who turned city-owned, foreclosed private buildings into cooperatives. By 2004 there were 1.5 million units of cooperative housing in the United States, but in some cities real estate appreciation was increasing the pressure for privatization.[64]

Whether access to affordable housing is provided through CDCs or cooperatives, there has been little interest in environmental goals or environmentally oriented design innovation. When I have asked housing activists about the issue, they say that there is so little money available that the primary focus is on providing housing, period. The dilemma is similar to that found in the other access pathways, where the primary focus is on access to conventional goods and services, and the more environmentally oriented designs are seen as luxury goods. However, there are some recent changes that hint at a greening process. Global Green USA, an affiliate of Green Cross International, has a Greening Affordable Housing Initiative that provides builders with low-cost design solutions that have significant payoffs in terms of building expense, long-term maintenance, and energy consumption. Some states and local governments have set up incentives for builders of green subsidized housing, and Habitat for Humanity has a greening initiative.[65]

To summarize localist and access action in the infrastructural field, the localist strategy of import substitution for bus and fuel purchases already has shown evidence of incorporation and transformation for the private-sector firms. For example, ethanol refineries have undergone consolidation, and import substitution for bus manufacturing appears to be risky when the firm attempts to grow beyond local and niche markets. There is also some evidence that cohousing in the United States has undergone professionalization, as architects have taken up the challenge of organizing projects, and in some cases cohousing has shifted to a condominium model in response to banks' rules for loan approval. There is little evidence for the incorporation and transformation of ecovillages, probably because they remain such a marginal activity and are so radically distinct from mainstream housing design.

Regarding the historical trends for access action, because transit agencies are generally governmental entities, the organizational effects of the devolution and privatization of welfare policies are much less visible. Access politics have tended to focus on helping low-income residents in central cities gain public transit access to jobs in the sprawling suburbs. In cities where the rail/bus divide is configured to favor rail at the expense of a deteriorating bus system that is heavily used by low-income ethnic minority groups, access action can emerge over investments in the bus system. In contrast, the history of CDCs provides a good example of the effects of welfare devolution and privatization on activist organizations. Although CDCs were themselves early examples of public-private partnerships and local control of federal access programs, they underwent changes from the goal of locally owned small-business development that served low-income neighborhoods to the construction of affordable housing. As non-profit organizations shifted from advocacy to service provisioning of affordable housing, they also underwent professionalization. Although professionalization increased resource availability, it led to a distancing of the organizational mission and structure from grassroots clients.

Regarding the effects on the general technological field of urban infrastructure and building design, the localist pathways described in this section (import substitution for public transit manufacturing and fuel, the development of infrastructure friendly to pedestrians and bicyclists, and the construction of cohousing and ecovillages) seem quite marginalized, at least in comparison with the growth and scale achieved to date for local agricultural networks, distributed energy generation, and the resale and credit union industries. Government mandates for biofuel consumption standards and in-state production quotas probably stand the most chance for achieving rapid growth, but the scaling up process has already entailed a transformation from locally owned refineries to ownership by non-local, publicly traded corporations. Of the localist pathways evidence of technological innovation can be found in the design of the electric circulator buses, bicycle lanes, and ecovillages. Regarding access pathways, transit activism in several cities has affected bus purchase decisions as well as the general tilt of public transit toward rail or buses, and budding interest in green affordable housing may produce some design innovations. Synergies across fields are also beginning to emerge as environmental issues in the infrastructural field translate into access issues in other fields. For example, to the extent that affordable housing meets green building standards, energy costs will be reduced.

Finance

In the financial field localist economic institutions recapture investments and consumer expenditures that otherwise would leave the region through the profits and overhead charges of non-local banks and retail chain stores. In this section I will consider three examples of localism in finance: credit unions, local currencies, and "buy local" campaigns. Although the locally owned commercial bank is increasingly rare, it could also be counted as a localist economic institution; however, it is not discussed here.[66]

Historically, credit unions have provided a significant alternative to commercial banking and one that has also tends to offer greater access to the less creditworthy. Unlike commercial banks credit unions are member-owned cooperatives that usually have some form of membership restriction, such as working for a specific firm (for a company-based credit union) or living in a delimited geographical area. Because credit unions are often geographically restricted by the membership criteria and are governed by the members, a shift in investment from a non-local commercial bank to a credit union is another example of import substitution.

Credit unions grew dramatically during the 1920s, when people were looking for inexpensive consumer credit and were unable to get consumer loans from commercial banks, and they underwent a second large wave of growth during the 1970s. The United States and Canada are the world centers of credit unions: as of 2002 the two countries had approximately 90 million members in 10,000 credit unions and about $478 billion in loans, compared with a worldwide total of 43,000 credit unions with 136 million members in 91 countries and $532 billion in loans. Historically, many of the credit unions in the United States helped the marginally creditworthy to secure home, automobile, and other kinds of loans. Credit unions have been a substantial source of financing for the working class and working poor, although generally not the very poor.[67]

At present the non-profit status of credit unions limits direct acquisition by commercial banks (a pattern that has been evident in other fields), but there is a strong pattern of consolidation. Deregulation in the financial services industry has allowed mergers and acquisitions in order to achieve economies of scale and reduce operating costs. From 1980 to 2004 the number of large credit unions (more than $1 billion in assets) grew steadily from 2 percent to 33 percent of total assets, whereas the number of small credit unions shrank by 50 percent. In New York State, the number of small credit unions declined by 29 percent during just the period 1999–2004, mostly through absorption into larger credit unions.

By 2006 the trend had become pronounced enough that the New York Credit Union League announced a plan to help the small credit unions survive.[68]

Although credit unions represent only about 7 percent of commercial banking assets, the commercial banking industry has come to regard as unfair competition the growth of large credit unions with relaxed membership restrictions, such as serving a multi-county area rather than restricting membership to the employees of one firm or a small town. In the wake of the state government budget deficits in the early 2000s, commercial banks mounted an offensive to convince legislatures to pass plans to tax large credit unions. The Credit Union National Association's 2002 annual report claimed that the commercial banking industry was attempting to take advantage of state budgetary shortfalls to put credit unions out of business and to divide them into some that were subject to taxation and some that were not. Battles between banking and credit union associations heated up in state legislatures and in Congress.[69]

A second type of localism in finance is community currency. There is a long history of local and regional currencies in the United States, but since the 1960s three models have become prominent. In the Ithaca Hours system, a local currency is printed and distributed among local businesses that agree to accept the currency. Although the model has not been widely replicated, it has been successful in the small city of Ithaca, New York, in part due to the civic culture (a progressive university town) and in part due to the substantial participation of the local business community. The much more extensively replicated model is the LETS. Sometimes defined as a "local exchange trading system," its founder insists that the original meaning of LETS was simply "let's." In a LETS, a group of people agree to set up an account of credits and debits that involves trading goods and services that they have available. The system was originally established for an economically depressed area in British Columbia where cash was in short supply, but it has since been explored under many different conditions. A third model is the local script, in which a business issues what are in effect interest-bearing coupons for future goods and services instead of seeking a bank loan. For example, a business will sell an IOU for $10 that is worth $11 at a future date if redeemed on the store's goods. Local scripts are similar to the import-substituting strategy of self-capitalization through trade and consignment found in the resale industry and the prepaid subscriptions of CSA farms.[70]

Some activists characterize their work as contributing to a local currency movement, and others describe it as a complementary currency movement. The latter term underscores the non-local quality of some of the innovations, such as Internet-based LETS systems among non-local communities and business scripts that are used across regional or even national markets. The term "complementarity" also reflects the prevailing view that the currencies occupy niches that serve particular economic roles, such as promoting local businesses or linking communities of interest, rather than replacing government-issued currencies. Even where the localist aspirations of the movement are paramount, the local currencies have to some degree become convertible due to the emergence of Internet-based exchange systems.[71]

A third type of localist action in the financial field involves substituting consumer expenditures from non-local retail outlets with purchases from local stores. In Boulder, Colorado, an alliance of 150 local businesses formed in 1997 to develop consumer awareness of the benefits of buying from locally owned retailers. The organization developed a logo and product labels to encourage consumers to spend their money in locally owned retail businesses rather than superstores. The organization also issued a community benefit card that provided discounts at about 60 local businesses, and it supported city government ordinances that would give preferences to local businesses and provide checks on the growth of chain stores. Subsequently, the American Independent Business Alliance was formed to develop similar projects in other cities, and by 2004 the alliance had grown to about two dozen affiliates. The organization has flourished mostly in small cities and college towns, that is, where the population is relatively homogeneous and perhaps more willing and able to support the value of locally owned businesses. Because the emphasis is on small-business development, sustainability issues are not a primary concern. And in some cases, small businesses do not want to be associated with local environmental politics, for fear of losing customers.[72]

An alternative local business alliance is the Business Alliance for Local Living Economies (BALLE), which grew out of the 2001 meeting of the Social Venture Network, a progressive business network. Unlike the

American Independent Business Alliance, BALLE membership is less focused on main street retail and has a strong social responsibility and sustainability agenda. Membership is limited to locally owned, privately held businesses. Definitions of "local" vary across the BALLE chapters from a small state to metropolitan regions to specific cities within a metropolitan region. Although member organizations do not have to pass a certification test or meet a standard, they must be willing to learn about how to implement environmental and social responsibility innovations in their businesses. Like the American Independent Business Alliance chapters some BALLE chapters have developed local first campaigns that encourage residents to shop in locally owned businesses. However, BALLE also sponsors educational meetings for members interested in rethinking business practices, and it is developing a solution to the problem of non-local localism: a website so that BALLE members in one city can make purchases from members in another city.[73]

Localist financial institutions discussed up to this point sometimes increase employment and credit opportunities for low-income residents, but they tend to serve the small-business sector, or, in the case of credit unions, the employed who have some physical collateral. Based on my interactions with people who have operated a LETS, some of the systems could even be characterized as a middle-class hobby, akin to domestic gardening and the home-repair projects of home-power enthusiasts. In contrast, a parallel set of financial institutions has developed with access for the very poor as the primary goal. Community-development financial institutions have been classified into five main organizational types: community-development banks, community-development loan funds, community-development venture-capital funds, community-development credit unions (CDCUs), and micro-enterprise lending. This section will focus on the last two categories: CDCUs, which provide mostly personal loans, and micro-enterprise finance, which is oriented toward small-business development for people who lack physical collateral. The section will also consider time banks as an access-oriented version of community currencies. As with other access institutions the primary orientation of CDCUs, time banking, and micro-enterprise finance is to provide financial or, in the case of time banking, quasi-financial resources to the poor, and consequently environmental issues are not a salient concern.[74]

CDCUs have a volatile history of ups and downs that reflects their situation as a political football in the broader game of welfare politics. During the 1960s the Office of Economic Opportunity provided funding for the establishment of low-income credit unions as part of President Lyndon Johnson's War on Poverty. However, many of the credit unions initiated during the period failed, in some cases due to poor planning and in other cases due to withdrawal of federal government subsidies. In 1970 the Credit Union National Association issued a report that was critical of the low-income credit union strategy. Defenders of low-income credit unions claimed that they had not been given a fair chance, and the credit union community became divided. In 1974, a breakoff organization, the National Federation of Community Development Credit Unions, was formed. The Carter administration soon expressed interest in CDCUs, and after a great deal of negotiation, a modified law was passed in 1979 that supported low-income credit unions.[75]

The victory was short-lived, because by 1982 the Reagan administration had cut funding for CDCUs as part of its overall strategy of cutbacks to anti-poverty programs. Many CDCUs went out of business, and the national federation suffered a budget collapse from the already meager sum of $500,000 to $5,000. In an effort to find new funding sources, the national federation began a grassroots project in New York to build new credit unions that would serve neighborhoods affected by the closure of commercial bank branches. By the mid 1980s the Reagan administration was looking for self-help approaches to poverty, and it began to support CDCUs as a candidate for private-sector funding without government financial support. Foundation support and loans from major banks continued through the 1990s, and the national federation was able to provide capital support to new credit unions. The Clinton administration's 1994 Community Development Financial Institution Fund brought new funding into CDCUs, and the number of low-income credit unions grew from 134 in 1991 to 419 in 1998. The national federation's membership grew to 215 member institutions in 2003, representing 700,000 members and $2 billion in assets, and the estranged relations with the Credit Union National Association improved. However, against the backdrop of the total membership and assets of credit unions in the United States (as high as $700 billion in 2006), let alone socially responsible investing as a whole (about $2 trillion) or commercial banks (about $4 trillion), community-development credit unions occupied a tiny portion of the market.[76]

A second access-oriented pathway in the financial field is time banking, which is essentially a community currency based on hours spent working for credit rather than the dollar value of goods and services traded. Started by attorney Edgar Cahn in the 1980s, the tax-exempt programs enroll the elderly, teenagers, and the unemployed, who undergo a role change from recipients to providers of support. When time bank members provide services such as shopping assistance, baby sitting, home care, and yard work, they accumulate time dollars that can be exchanged for other services. The systems are like LETS, but the exchanges are based on everyone receiving the same amount of credit for one hour of labor, and the focus is on poverty reduction and community building. In the process, civil-society institutions and community networks are strengthened.[77]

Time-banking programs have spread throughout the United States, Canada, and the United Kingdom, and in some American states welfare recipients may participate without losing benefits. Although time banks have been proven successful, comparative case study analyses indicate that they face several significant obstacles, of which the most pressing is long-term financial self-sufficiency. Usually the time-banking programs require an office and staff in order to recruit individuals and community organizations. Although foundations provide startup funds, like the micro-enterprise programs time banks face a long-term challenge after the initial funding runs out. One solution is to find a place under the wing of a larger, more established organization within the community, such as a community bank. Another significant obstacle has been that people tend to give time and accumulate credits without spending them. The remedy is education and diversification of members, so that there is a wide enough range of services for people to find ways to spend their credits, but the remedy requires additional investments of resources and management. In some cases people may also give their credits to others.[78]

A third access-oriented pathway is micro-enterprise development finance, which consists of loans and training programs for the very poor to help them to develop businesses. The model was developed in Asia and Latin America before being imported into the United States during the 1980s. One of the most powerful institutional forms for micro-enterprise finance is the solidarity group model, which was developed in the non-profit Grameen Bank of Bangladesh and ACCION International. The Grameen version of the model involves sending out field representatives who go from door to door, actively recruiting potential borrowers from the poorest of the poor (usually women). The goal is to provide small loans, usually under $50, to help women start home businesses or to help them break cycles of usurious borrowing in their current micro-enterprises. Borrowers go through extensive training and form a solidarity group with other borrowers, who develop group savings by slowly accumulating a portion of the revenue that is set aside. Because loans are granted to the solidarity groups rather than individual borrowers, there is significant peer pressure on individuals to repay the loan. The active role of the bank in recruiting, training, and overseeing the solidarity groups results in high overhead rates, but in compensation the default rate is low.[79]

By 2002 there were about 7,000 NGOs in the world providing micro-enterprise financing, and the Microcredit Summit Campaign of 2003 announced a goal of reaching 100 million of the world's poor. Notwithstanding the successes, during the 1990s microfinance in low-income countries underwent a rapid commercialization process that resulted in a more diverse field of institutions. A few NGOs became commercial financial institutions in order to gain access to savings and higher levels of capital. At the same time, credit unions entered into micro-enterprise financing in some areas, and some commercial banks moved downmarket in search of new opportunities. Commercialization provided access to higher levels of credit, but it tended to be accompanied by a shift in loan strategy from solidarity groups to individuals with personal collateral. Furthermore, by foregoing the training programs, supervision, and solidarity group checks, the commercial bank programs had higher default rates.[80]

Micro-enterprise finance in the United States has operated under very different economic and political conditions than in the low-income countries. With the decline of welfare programs since the Reagan administration, micro-enterprise finance became an attractive alternative to welfare for both Republicans and third-way Democrats. Clinton supported microfinance as governor of Arkansas and continued to do so as president, but there had also been support for micro-enterprise finance during the administrations of Presidents Reagan and Bush. Even before their presidencies the Community Reinvestment Act of 1977 under President Carter had allowed banks to gain community reinvestment credits by providing loans to microfinance organizations. Other sources for the development of micro-enterprise finance in the United States came from the feminist movement, which supported some of the early micro-enterprise finance programs, and CDCs, which started many of the programs. By 1991 the field was developed enough that the organizations had formed a trade association, and in 1992 the Small Business Administration inaugurated a microfinance program. The number of micro-enterprise finance organizations increased from about 100 in 1992 to more than 400 in 2002.[81]

In the United States, government regulations and competitive business markets have made it difficult for the micro-enterprise finance programs to flourish in the same way that they have in low-income countries. Low-income American borrowers needed more business training, and there are fewer opportunities for entrepreneurial small businesses because of the lower self-employment rate. Micro-enterprise finance programs also run into conflict with welfare laws; for example, in Illinois the Full Circle Fund had to receive special permission from the welfare authority before it could approach welfare recipients, and eventually the program requested and achieved a change in the state law. Even the 1996 welfare reform legislation that was championed by President Clinton, a supporter of micro-enterprise finance, introduced work requirements that ran into conflict with attempts to pursue self-employment through micro-enterprise loans.[82]

In the United States there is little evidence of the commercialization pattern found in low-income countries, where non-profit organizations have shifted into commercial banking, and commercial banks have moved downmarket. Credit is more widely available in the U.S., where loans and credit from credit cards are available to people who have a steady job history and some physical collateral. By targeting borrowers who otherwise would not have access to credit, American micro-enterprise finance programs have found it necessary to invest resources in skill development so that the loans are successful, and consequently they have tended to shift toward training and away from lending. The shift can be justified from a poverty reduction perspective, but it increases overhead costs and makes financial self-sufficiency more difficult for the micro-enterprise organization. The shift also puts the organization in a bind, because funders such as foundations want to see high levels of lending and long-term organizational financial self-sufficiency. To achieve those goals, micro-enterprise organizations must move up the income scale to less severely poverty-stricken clients, frame their work in more businesslike language, and professionalize the staff. As a result there is a trend that is parallel to the commercialization process in low-income countries, but it tends to occur within the non-profit organizations, analogous to the professionalization process within CDCs in the United States.[83]

To summarize the general historical trend of localist and access action in the financial field, the non-profit status of credit unions and local currency organizations has allowed them to continue as localist entities and avoid consolidation into large, publicly traded corporations. Consolidation has occurred among some of the smaller credit unions (but into larger credit unions), and if credit unions were to become taxable, it is possible that some would undergo organizational changes that would ultimately make them amenable to acquisition by commercial banks. Another potential trend is the trading of community currencies on electronic auction sites (analogous to the conversion of the local yard sale economy to global trade), but it is unclear how much potential there is for such trading to undermine community currencies.

Regarding the three types of access-oriented alternative pathways in the financial field, there is evidence of significant historical change since the 1960s. CDCUs were affected by cutbacks in welfare programs, but because they were already in a service-provisioning mode, the main effect of the cutbacks was to spur greater private-sector partnerships with commercial banks rather than to force a transition from activism to service provisioning. Likewise, micro-enterprise finance programs were already in a service-provisioning mode, and for them the trend has been increasingly towards an emphasis on education and skills training under the guidance of a professionalized staff. In contrast, the primary change (and challenge) of time banks has been to seek long-term financial stability by finding support under the wing of a non-profit organization.

Regarding the effect on a technological field, the financial products offered by credit unions are comparable to those offered by commercial banks, but the prices and the terms may be more advantageous to borrowers and savers. In contrast, the financial products offered under micro-enterprise loans are innovations to the degree that they include features such as lending circles and training. Clearly, complementary currencies and time banks are different financial products from conventional currencies, but their general effect on the economic well-being of local economies, small businesses, and the poor remains to be demonstrated. Although business-based scripts have some potential for self-capitalization that extends the model beyond the subscription agriculture idea and store credit trading in the resale industry, the principle effect of LETS and community currencies may end up being their use as advertising instruments that draw attention to local businesses. The idea of supporting local businesses by shopping locally appears to be gaining ground, especially in the wake of rising criticism of the labor and environmental practices of some of the superstore chains, but the fundamental problem of price competition between retail outlets that rely on wholesalers and those than do not is inescapable. As a result the locally owned retail industry will have to continue to differentiate itself through innovation, such as by offering higher quality products and services and by drawing attention to the localist provenance of products.

Conclusions

As the various cases in this chapter have underscored, localist and access pathways are not necessarily green, but at some points there are connections with environmental values. Localist pathways exhibit evidence of concern with environmental issues in local agricultural networks, some public power agencies, home-power projects, reuse centers, alternative fuels in public transit, local building materials, cohousing, ecovillages, and the BALLE business networks. However, the point should not be overstated: locally owned farms that sell in farmers' markets are not always organic; some of the greenest public power agencies are currently building natural gas plants; the most militantly off-grid home-power projects can include propane gas for kitchen stoves and wood-burning fireplaces for home heating; cohousing units can be built with conventional construction materials and follow conventional, condominium-style design; and locally owned, independent, Main Street, retail businesses are sometimes afraid that becoming too openly green may alienate their more conservative customers. Nevertheless, to the extent that the substitution of local production and products with distant ones creates awareness of environmental externalities and also reduces the energy costs of transportation for product distribution, localism can have environmental dividends.

As with localism, the access pathways exhibit some points of convergence with environmental concerns: community gardening tends to use organic and quasi-organic horticultural methods; some utilities offer low-income weatherization programs in their low-income programs; thrift stores divert clothing and other household goods from landfills; affordable housing and public transit are being offered in less toxic and more energy efficient forms; and micro-enterprise and credit union loans can support import-substituting green businesses. Again, the connection should not be overstated. The primary concern of access action is to bring resources to people who lack adequate food, energy, clothing, household goods, housing, transit, and credit. With a few exceptions, the access pathways are not a site for technological innovation, including green design. For the hungry the central issue is securing food. It would be nice to have high-quality, fresh, organic, locally grown food, but the overburdened food banks and pantries must take what they can get.

Just as neither localist nor access pathways are not necessarily concerned with environmental values, so localist pathways do not necessarily show great concern with access for the poor, and access organizations do not always show great concern with local economic control. There is some evidence that localist pathways can address access issues for low-income residents, but it is mixed. Examples include the operation of farmers' markets in low-income neighborhoods, scholarships for community-supported farms, job training programs and affordable home furnishings found in the non-profit reuse centers, the extension and improvement of public transit, the emergence of urban ecovillages, and living wages advocated by progressive small-business alliances such as BALLE. However, there is no necessary connection between localist institutions and access goals; in fact, much of the activity in local agricultural networks, home power, cohousing, ecovillages, credit unions, community currencies, and progressive small businesses has a middle-class social address. Localism is not a poor people's movement; to the extent that it can be viewed as a movement, it is rooted in the small-business sector's opposition to globalization and, in sometimes uncomfortable coalitions, in the countercultural lifestyle movements of relatively affluent green consumers.

In a similar way, the politics of access do, in some cases, intersect with those of local economic control. Block grants to state and local governments, encouragement of public-private partnerships, the rise of faith-based service-provisioning organizations, conversion of welfare to workfare, and general cutbacks in both federal programs and block grants have become abiding features of the neoliberal political landscape of the late twentieth century and the early years of the twenty-first. What liberals saw as the federal government's responsibilities to the poorest members of society have been shifted downward to the local government and outward to civil-society and private-sector partners. By default access organizations operate increasingly at a local level and must grapple with local economic development, which can include the development of small businesses and partnerships with localist institutions.

Regarding the historical patterns of change in both localist and access movements, it is now possible to draw together some observations in the form of preliminary hypotheses regarding the particular form of incorporation into the agendas of mainstream industries and government agencies, and how the original goals of organizations and broader pathways of change can be transformed. For the localist pathways there is a tendency toward the consolidation of private-sector firms into larger, national franchises or corporate retail chains, as occurred in the example of sustainable farms that entered food processing and in the for-profit retail resale industry. In contrast with firm-based localism, non-profit and voluntary organizations are less likely to be displaced or incorporated into national for-profit chains and franchises. However, they tend to undergo professionalization, formalization, and restructuring of the board, and in the process they become open to non-local influence and even non-local control. Furthermore, regulatory policies that encourage privatization, such as policies that would require public power agencies to divest from ownership of energy generation or mandate that credit unions pay taxes, could significantly weaken public and non-profit localist institutions.

The relationship between the Internet and localism remains open to development and further research. On the one hand, the Internet allows small firms and even households to operate in global markets at a relatively low cost. With appropriate labeling the Internet could enhance the development of non-local localism, that is, the sale of goods produced by localist institutions in continental and international markets. An example is the plan for BALLE chapters to develop Internet-based business-to-business networks. On the other hand, the Internet undermines localism by providing non-local, small-business competition for local, small businesses much in the same way that "big box" superstores undermine locally owned small businesses. To combat the risk, there would have to be a strong consumer preference in the order of local businesses first, followed by non-local small businesses and then corporate retailers. There is also some potential for localist institutions such as community currencies and yard sales to be undermined by resale and trading on the Internet.

Another generalization is that the import-substitution strategy of localism appears to work more easily under some conditions and in some industries. For example, when a transit agency can displace purchases from distant corporations to local biofuel refiners and local bus manufacturers, there will be little opposition as long as the changes do not cause so much financial hardship that service will be reduced or prices raised. In contrast, the price and lifestyle premia associated with investing in distributed, renewable energy or moving into a cohousing facility will probably condemn those forms of localism to marginal status, unless there are government-sponsored incentives based on increasing concerns with greenhouse-gas reduction and energy conservation. Localism probably will be most successful where there is a direct economic benefit (such as cleaner fuel or better access to credit), the price premium is associated with higher quality (such as local, organic food), and the shift in purchasing preferences is relatively easy to accomplish (such as consumer goods that are readily available in locally owned retail outlets).[84]

A general weakness of localism is that to date there is no evidence of a solution to the problem of providing consumers with choices for non-local manufactured goods. Although there are fair-trade shops for gifts and specialty items, and it is also possible to buy many manufactured consumer goods through the resale market, in the long term localist institutions operate on the fringes of the global consumer economy. To make localism work in a global economy, consumers would have to be able to buy (as a second choice when local sources are unavailable) a non-local product that has been made with fair labor practices by a community-oriented business that is not publicly traded, such as a cooperative or a small employee-owned manufacturing firm. Product labeling exists in incipient form in some of the foods that are sold with double certification as fair-trade and organic. However, to expand the potential into the realm of manufactured consumer goods seems to fly in the face of all economic trends. Given the differential of wages at a global level, small American manufacturing operations cannot compete unless they have a unique product that is protected in some way, such as by a patent, or it can be manufactured in a capital-intensive manner that can compensate for higher labor costs. We do not yet know what percentage of consumers would be willing to pay a premium, or how much, for a doubly certified manufactured good such as a lamp or a computer. As a result, I believe that localism and the technopole should be viewed as complementary regional development strategies that should be explored in an integrated way.

Organizations that would like to take localism to the next level—such as a global trade in manufactured products that are doubly certified as environmentally responsible and produced by community-oriented, independent businesses and non-profits—would also require access to capital to achieve a manufacturing scale and ongoing cycle of product innovation that would allow them to be competitive globally, even with a price premium. However, the primary way to access capital markets is through sale of equity, and when equity is sold local ownership and control are often forfeited. Some of the leaders of BALLE are examining the problem of localism and capitalization. One possibility is the development of local stock markets, which would be limited to buyers who live a geographical area. However, such alternative institutions would also have to be combined with alternative corporate charters that ensure that sustainability and fair labor practices are paramount goals that supercede the race for short-term earnings maximization; otherwise, they will have merely reinvented the wheel, or the treadmill, at a smaller scale.[85]

Whereas in the localist pathways the pattern of incorporation and transformation is direct in the case of for-profit firms and still in incipient stages for firms that want to sell in global markets, in the access pathways the pattern of incorporation and transformation is more closely tied to changes in non-profit organizational missions in response to government policies. During the 1960s there was more evidence for social-movement activity. At the time the federal government recognized its responsibility to the poor through entitlement programs, and anti-poverty movement organizations sometimes won policy changes from the federal government. However, the design of anti-poverty programs tended to reproduce relations of dependency rather than enhance political participation and economic power from below. Beginning with the Nixon administration and, with a partial hiatus during the Carter administration, continuing through subsequent administrations, the devolution and privatization of the welfare state has tended to shift activism and advocacy politics to the local level and transform organizational missions into a service-provisioning role. The result has been a depoliticization of the movement dimension of the access pathways and a return to charitable activity that has historically rested with faith-based organizations. New institutions that have emerged in the wake of the decline of federal entitlement programs (food banks, energy banks, furniture banks, CDCs, CDCUs, micro-enterprise loan programs, and so on) as well as newly reinvigorated old institutions, such as community gardens and non-monetary exchange systems, have helped redefine the politics of access as, largely by default, locally oriented and in some cases locally controlled.[86]

As organizational missions in some cases shifted toward service provisioning, organizational structures also tended to change. Volunteer, staff-based direction tended to give way to formal, non-profit corporations with external boards that were selected to help provide access to extramural funding through partnerships with corporations, foundations, faith-based organizations, and government block grants. Partnership is the primary route of incorporation of access-based institutions into the world of corporate control, in contrast with the more direct routes of consolidation and acquisition for the private-sector firms of the entrepreneurial side of the TPMs and the localist pathways. There are some exceptions, such as the entry of for-profit firms into the thrift industry, but in general the non-profit organizational structure of the access field has generated a somewhat different process through which elites exert control over grassroots organizations. Although there is a trend toward service provisioning and professionalization, even service-provisioning organizations with externally oriented boards of directors have a capacity to mobilize, not only with respect to local governments but also through national umbrella organizations. The transformation from advocacy to service provisioning can lead to second-order advocacy, as service-provisioning organizations jump back toward advocacy when threats to government funding loom. One example is the role of networks of fuel banks, which have supported attempts to stop legislative efforts aimed at additional cutbacks in block grants or other forms of federal government assistance.

On the issue of innovation and the effect on a technological field, the localist and access pathways tend to complement the IOMs and the TPMs. Whereas technological innovation and the design of technologies and products is crucial to understanding the politics of IOMs and TPMs, organizational innovation is more prominent among the localist and access pathways. Completely new types of organizations have been invented, and existing ones, such as cooperatives, have been redeployed. New mechanisms of financing (including public-private partnerships, subscription agriculture, self-capitalization of retail stores through trading in used goods, and solidarity groups) should also be included among the organizational innovations. In some cases the localist pathways demonstrate substantial technological innovation, such as in distributed renewable energy, building deconstruction, ecovillage design, and complementary currencies. However, other than work by some social scientists, there are generally no scientific research fields associated with the localist designs of alternative technologies, products, and infrastructure; instead, the knowledge is developed in the informal networks of advocates and activists and circulated in non-academic conferences. Their knowledge production and their technological innovations emerge outside the world of patents and publicly traded organizations, and they remain largely invisible to policy makers and university-based researchers. However important such institutions may be for thinking through and implementing a vision of a society that is not only more sustainable but more just, we know very little about their potential and their shortcomings. The study of localist and access pathways is largely a walk through the fields of undone science.[87]

Conclusion

The commonsense understanding of innovation usually begins with a scientific or industrial laboratory where researchers develop new ideas, sometimes with industrial funding, that are potentially of value to society and to industry. A technology transfer office obtains patents; the patented knowledge is then licensed to a business, which eventually brings a new product to market. The recognized role for democratic input into the process of innovation is limited to three points: voting for elected government representatives, who can influence budgetary priorities for scientific and industrial research; a decision by a regulatory agency, which has been delegated the responsibility of examining new products from the perspective of social and environmental welfare; and a purchase by a consumer, who selects one product over another.

Under conditions that have variously been described as globalization and late capitalism, there is little room for democratic participation in the governance of industrial innovation. Congressional elections are heavily influenced by corporate spending, regulatory agencies have restricted budgets and are often partially or completely captured by the regulated industry, and consumers' decisions are heavily influenced by product placement, advertising, and public relations. A number of solutions have been proposed to enhance democratic oversight of scientific research and technological innovation, and there is some evidence that public participation has been broadened by institutional changes such as diversified social composition in the scientific and technical professions, increased attention to community-based and participatory research, and the growth of participatory and user-centered design. Such reforms are valuable and warrant further exploration, but they often allow public participation only in an advisory capacity and presume that people who are engaged in research and development have the good will to take seriously the suggestions from people who are not. As I have argued, social movements, reform movements, activist networks, advocacy organizations, and other alternative pathways for social change can also enhance democratic participation in shaping scientific research fields, technological innovation, and industrial change. The alternative pathways can also be sources of organizational and technological innovation.[1]

Incorporation and Transformation

Social scientists and historians have long studied the cooptation and routinization of social movements, and much of my analysis can be seen as a contribution to that research tradition from the perspective of science and technology studies. Agents of social change often find, to their chagrin, that they have made history, but not exactly according to their original vision. Rather than achieving a full victory, they usually become caught up in a more complex dance of partial success and cooptation. In this sense my examination of the incorporation and transformation process of the alternative pathways develops a well-trodden intellectual terrain. However, my focus on material culture and industrial change, and how the processes of partial acceptance by elites play out in the definitional struggles that I have termed "object conflicts," is arguably more novel, and it is one indication of how the study of science and technology might profitably be conjoined with the study of social movements. In the examples that I have discussed, elites can be found incorporating and transforming the policy changes that social movements advocate (a process that is well recognized in the general literature on social movements), but they can also be found remaking knowledge, technology, products, and the organizational relationships in ways that respond to the action of social movements.

The incorporation and transformation of social-change action into state policies and industrial priorities has parallel but different patterns across industrial opposition movements, technology- and product-oriented movements, and localist and access pathways. In IOMs, incorporation takes place largely through the politics of the partial moratorium, that is, the achievement of an end to a particular aspect of sociotechnical system. On the negative side, the achievement of a partial moratorium often takes the wind from the sails of a movement; on the positive side, it creates a new ground upon which subsequent generations of mobilization will be based. In TPMs the process occurs largely through the uptake of technology and product innovations by a targeted industry. As the mainstream industry shifts from resistance to incorporation, the companies may acquire the innovating entrepreneurial firms or develop new product lines, and they often redesign alternative technologies and products so that they become complementary to existing production technologies and products. As the process becomes advanced, object conflicts or definitional struggles, especially between the more alternative and complementary versions of the innovation, intensify. Regarding the localist pathways, the process involves the allure of distant markets and the threat of non-local control and ownership. The allure and threat are particularly salient for private-sector firms (such as alternative, locally oriented farms and resale businesses), but non-profit organizations also undergo a transition from staff control to board control. In several fields, access-oriented pathways also undergo a long-term process of incorporation and transformation, as advocacy and social-change goals shift to service provisioning under conditions of welfare devolution and privatization.

From an optimistic perspective, the world is a somewhat different and better place after the incorporation and transformation of the social-change action. Societies change as a result of social-change action, albeit not enough from the perspective of the challengers and too much from the perspective of the elites. The alternative pathways create a politics not so much of creative destruction (in the Schumpeterian sense) as of creative reconstruction. After incorporation and transformation have taken place, the various political, scientific, technological, and industrial fields are different, and social-change organizations regroup and develop new priorities and campaigns. They begin again with a new set of political opportunities and constraints; the historical field for action has shifted. In this sense the movements do matter.[2]

From a pessimistic perspective, the alternative pathways undergo a process of cooptation and, in some cases, division. To the extent that the alternative pathways seek fundamental transformations in patterns of ownership and economic control, they are generally not very successful.

We might call this the Pynchon principle: you can tickle the master's creatures, but you cannot touch the master. For example, the idea that the demise of nuclear energy (or of fossil fuels today) would coincide with the democratization of energy through locally controlled renewable energy production is not how the history played itself out. A partial moratorium on nuclear energy production was achieved, though it was due to changing economics as much as to social-movement mobilization, and a partial greening of the power industry has occurred. However, the fundamental ownership patterns and growth in aggregate consumption and greenhouse-gas generation continue unabated, and the transition to renewable energy is far from the dreams of the original activists, let alone at levels that scientists now say are necessary to halt the trend toward global warming. From a pessimistic perspective, the movements affect a distributive politics within elites that shift advantages from one industry to another. When movements are oriented more toward ownership issues, such as the mobilizations against the World Trade Organization, they are on a long road. In contrast, the more technically oriented changes of industrial priorities, such as shifting corporate investment priorities away from rainforest timber harvesting, sometimes show higher levels of success.[3]

In addition to bringing about partial and modest changes in society, from an optimistic perspective the alternative pathways play another historical role. Taken as a whole, the diverse alternative pathways articulate a vision of a better society that challenges the prevailing ideology of neoliberal globalism and encourages a broader debate about what a society based on the values of justice and sustainability might look like. It is here that the voice of the social scientist (the analyst of what has been and is) ends, and that of the philosopher and citizen (the normative inquirer into the just and sustainable society) begins.

Sustainability and Justice

As a whole, the alternative pathways articulate a vision of how to transform American society (and, presumably, other industrialized societies) in a more just and sustainable direction. I have intentionally included the access and localist pathways, which articulate issues of justice more directly than IOMs and TPMs, in order to avoid a common

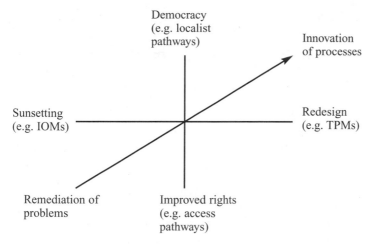

Justice and sustainability as a field of contestation.

shortcoming of discussions of sustainability, which can be subject to a double emptying process. On the one hand, the term "sustainability" is disconnected from a concern with justice, and as a result scholars and activists must remind general audiences of the connection. On the other hand, the concept is reduced to the mildest forms of the greening of production and consumption, in which an incremental environmental change or remediation of some of the most egregious environmental practices can be an occasion for the "greenwashing" of corporate and government reputations. Because of the double emptying process, we need a concept of sustainability that includes justice and can accommodate a range of technological and organizational innovation.[4]

The concept of "just sustainability," as developed by Julian Agyeman and his colleagues, provides an explicit alternative to the technocratic versions of sustainable development that have become popular in some policy circles and that have made a discourse of sustainability safe for ongoing environmental destruction and increasing inequality. I suggest that the various epistemic, technological, and organizational innovations of the full range of alternative pathways discussed here provide an empirically anchored vision of a justly sustainable society. The vision is consistent with that of Agyeman and his colleagues, and it is in the tradition of those who advocate a red-green or brown-green understanding of sustainability as requiring concurrent consideration of justice issues.[5]

I have integrated the four types of alternative pathways discussed in this book into a diagram of just sustainability as a field of contested values and action. (See the accompanying figure.) The vertical axis refers to building just communities (sometimes called the "red" or "brown" dimension); the horizontal axis refers to industrial and technological greening (sometimes called the "green" dimension). The concept of a justly sustainable society is normative; I assume that moving in the direction of a justly sustainable society represents societal and ecological progress and that the upper right quadrant represents a deeper understanding of just sustainability than the lower left quadrant. The value of progress articulated in the figure informs the choice of alternative pathways as a topic worthy of social scientific study, and it also informs the focus of that study on the question of the trajectory and outcome of the alternative pathways.

As the access pathways suggest, at the minimum a just society must remedy gaps in human rights, such as the absence of access to health care, food, clothing, and shelter, and the absence of basic political rights, such as open elections and protections on civil liberties. However, more fundamental economic and political rights, such as access to a livelihood with a fair wage and to participation in the political system, are often in need of remediation as well. Although the United Nations Declaration of Human Rights is a good starting point for a broad definition of human rights that goes beyond a purely political list of freedoms, it is not possible to enumerate a fundamental list of human rights on a permanent basis. The concept of rights constantly undergoes expansion to include new domains of social life, and likewise the innovation of new technologies and products creates an ever expanding field of shortfalls in minimal access to goods and protections from risks.

Even if one leaves the value of human rights open to expansion and development, the concept of a just society should include more than the remediation of rights. The other end of the axis points to the process-oriented goal of changing economic and political institutions to enhance democracy. Again, the meaning of democracy varies over time. Whereas the extension of the franchise was more salient in the past, increasingly the problem of how to protect the state at all levels from control by large, publicly traded corporations has become a more pressing institutional goal for the preservation of, or achievement of, democracy. Furthermore, today democracy is needed up and down the scale, and not just at the global level, where there is a recognized democracy deficit in global financial institutions, but also at the local level, where corporate globalization has resulted in significant and unwanted dislocations. There are many strategies for changing economic and political institutions to enhance democracy in an era of corporate globalization; among them is the development of alternatives to publicly traded corporations, restrictions on their economic influence over states, enhanced deliberative institutions and processes, and the development of locally owned economic institutions that are based on values of justice and sustainability.

The horizontal axis of industrial greening—that is, change in knowledge, technology, production, and consumption—is defined to suggest a parallel range of reform projects. As the goals of the IOMs suggest, there is great value in identifying existing environmental problems, such as unnecessary waste and the hazards and risks of industrial production, and to place a moratorium on the worst technologies and practices. The sunsetting of the bads of industrial society is often accomplished on a piecemeal basis through the partial moratorium, just as redressing violations of human rights often occurs through piecemeal measures such as raising the minimum wage or extending health care. Incrementalism may be the only feasible solution within existing political institutions, and it can result in significant improvements in the lives of some people and in the environment. However, incremental remediation can also take the steam out of efforts to achieve more profound change.

As the TPMs suggest, industrial change that remediates environmental problems can be distinguished analytically from a processual approach that examines more deeply design issues across the product life cycle. In practice there is a continuum between the two ends of the axis: by declaring a moratorium on unwanted practices and products, regulatory agencies motivate technological innovation. The range of industrial changes can involve substitution of raw materials as sources to products, new types of product usage, the recapturing of waste as inputs, and the use of renewable non-greenhouse-gas-generating energies to power some or all phases of the product life cycle. Many organizations that advocate and practice ecological design have a processual approach; among them are advocates of zero waste, industrial ecology, biomimicry, cradle-to-cradle design, closed-loop manufacturing, and living machines. When combined with zero emissions and a renewable energy source, those approaches to ecological design best articulate the concept of "redesign" as environmentally oriented innovation. Achieving sustainable product life cycles across industries requires significant changes in the design of production technologies, products, consumption patterns, energy sources, and disposal patterns, but it is also likely to involve changes in regulatory policies, investment norms, consumer preferences, and organizational missions. Ideally, a fully sustainable product would have to be created from waste, made and used with minimal or no waste, and be disposed as an input into a new product, a process that occurs in life itself and in human production based on biomimicry.[6]

Shifting toward a more just and sustainable society and especially building institutions that strengthen the more processual ends of the continuum will require changes in political culture and technology along with changes in organizational structures and institutional processes. Societies based on publicly traded corporations that must produce short-term earnings growth for stockholders have the financial resources to remediate human rights problems and environmental problems, but they are unlikely to engage voluntarily in the more profound changes that are characteristic of the processual ends of the field. In order to achieve a more just and sustainable society, fundamental changes in the basic structure of the private sector will probably be necessary.

A "Civil-Society Society"

In order for a society to be more just and sustainable in the full sense outlined above, the private sector would operate in ways that are currently more closely approximated by civil society. Specifically, a portion of the private sector, the publicly traded corporation, would have to be recast to resemble more closely the portion of civil society that operates as non-profit organizations and produces goods and services within an overall mission of environmental or social justice values. In a sense the productive function of society would be transformed so that it would filled by organizations that more closely resemble progressive small businesses and non-profit organizations.[7]

The demise of the publicly traded corporation as the dominant economic institution of modern society may seem to be a remote possibility, but one should remember that the institution as we know it today is a relatively recent phenomenon. Although there are antecedents in the more distant past, many of the primary institutional features of the modern corporation date back only to the railroad companies of the nineteenth century. Furthermore, from a long-term historical perspective a shift in sectoral dominance in society would not be unprecedented. In the small-scale, non-state societies that were once studied by cultural and social anthropologists, the kinship system governed the other societal sectors, whereas in the urban states and large empires based on warfare the political-military system became the dominant societal sector. If our modern societies were to evolve toward a future, post-capitalist type of society that could be capable of achieving a goal of just sustainability, it is possible, and arguably necessary, that a new societal sector would achieve dominance. Civil society has developed in the interstices of the capitalist order much in the same way that capitalist firms emerged from the interstices of the medieval order. The modern non-profit organization that produces goods and services from a values orientation of just sustainability provides a model of a way to organize economic production that takes advantage of the efficiencies of markets but embeds them in values that are necessary for ecological survival and social stability.[8]

A shift in sectoral dominance would not necessarily require the extinction of the publicly traded corporation and its replacement by a type of non-profit organization that produces goods and services. The publicly traded corporation could be redefined so that the goal of earnings growth for shareholders would take place inside a broader social and environmental mission that would be required of publicly traded entities. A "civil-society society" would exist if a convergence were to emerge between the private sector's role of producing goods and services and civil society's role of providing general societal and environmental benefits on a non-profit, voluntary basis. The convergence would occur in organizations that encompass the production of goods and services in a broader mission oriented toward the justly sustainable society. From an ecological perspective such convergence holds out the potential to improve adaptation by shifting the economy away from organizations that emphasize continued growth that drive unsustainable levels of consumption and ecosystem damage.[9]

The concept of a "civil-society society" provides a much more realistic and achievable goal for societal organization than the nineteenth-century vision of a post-capitalist future represented by socialism and the twentieth-century record of corruption and environmental degradation by government-owned industrial corporations. However, the concept should not be confused with the prediction that achieving it is a likely future trajectory for human history. If the record of injustice and environmental destruction of human history is any guide to the future, it is very likely that elites will resist attempts to make fundamental changes that could affect ownership patterns and profitability. The present trends represented by the rubric of "globalization" suggest that an alternative and rather bleak future is also a possibility: the increasing commercialization of science, the rise of agricultural biotechnology and ever longer commodity chains, the demise of natural gas and oil that may precede a return to nuclear energy and coal with much higher levels of pollution and climate instability, the development of a new generation of nanotechnological chemicals with unknown environmental and health consequences, continued concessions to sprawl and urban growth machines, and the dismantling of labor and environmental laws through global financial institutions that lack even the most basic democratic inputs.

If one steps back 100 or 200 years, and if one projects as far forward, the longer-term historical trends provide some grounds for pessimism regarding the potential for the vision of a more just and sustainable society to be realized. The societal trend has been for governments and markets to be increasingly internationalized and dominated by multinational corporations. Likewise, the ecological trend has been for the growth in ecosystem withdrawals and deposits to outpace increased technological efficiencies in production and consumption. If the trends were to continue, the resulting ecological crises would likely be managed through increased warfare, starvation, disasters, epidemics, and injustices of many other forms. The wealthy might be able to escape the worst aspects of future trends by living in artificial, enclosed environments that provide high levels of security, filtered clean air and water, non-toxic food grown in greenhouses, and high access to global commodity chains, whereas the poor could be condemned to live in toxic and insecure environments, to survive on the basis of localized and subsistence-based production and consumption, and to lack minimal access to global markets.

If the divisions between rich and poor were to become deeper and more widespread, then political instability would probably increase. In turn, elites could continue to endorse higher levels of surveillance and ever-weakening democratic institutions.

However, there are also some more optimistic trends that point to the possibility of achieving a different future scenario. The localist and access pathways provide examples of organizations that engage in the production of goods and services but also have missions of improved social justice and/or sustainability. An example is a privately held or non-profit farm organization that combines food production and sales with a variety of other functions, such as education, charitable food donations, and land stewardship through sustainable agricultural techniques. Another example is the privately held or non-profit reuse organization that combines an environmental and poverty alleviation mission. Such organizations are not concerned directly with increasing profitability or with growth, as is the mandate for publicly traded corporations. However, the organizations do not rely entirely on voluntary contributions to remain financially self-sufficient, as is the case with ideal typical civil-society organizations. Rather, the goal is for the organization to generate enough revenue through a combination of sales and voluntary contributions to remain in existence and propagate its social mission. Under such conditions, which could be described as production for reproduction, the organization is producing products, and some of those products are for sale in markets, but it may also solicit donations in the form of labor, cash or in-kind contributions, grants, or consumers' acceptance of price premiums. In other words, the paramount goals are developing a particular type of product via a particular type of production process, rather than production for profit.[10]

From the perspective of IOMs and TPMs, the social and environmental responsibility goals of large corporations in existing industries have yet to be emancipated from profitability goals. Some publicly traded corporations have taken a few steps toward social and environmental responsibility by endorsing codes of conduct and product certification standards. If the corporation could be protected by a label that certifies its products to be produced by fair labor standards and sound environmental standards, and if the corporation could attract the investment dollars of socially responsible investors and the consumption dollars of socially responsible consumers to make the shift economically feasible, then it may be able to resist market pressures to undermine those standards over the long term. For publicly traded corporations, much will depend on development of government regulations, continued oversight by civil society, redefinition of mission, expansion of responsibility codes, and increasing demands for socially responsible investment and consumption. In the current regulatory context such goals have been difficult to achieve over the long term, and in the absence of regulatory changes some corporations have decided that the best way to pursue such goals is to remain private or to go private by repurchasing stock from shareholders.[11]

One can select from current trends many other examples that would justify a relatively pessimistic or optimistic future scenario. It is not possible to predict whether human society will be more hierarchical and ecologically destructive 100 or 200 years from now, whether resource-based warfare will lead to civilizational collapse, or whether fundamental institutional changes in the direction of just sustainability will be widespread. Although a pessimistic scenario is valuable to hold in check optimism that can quickly shade into a utopianism that ends in disillusionment, developing the more optimistic scenario of an emergent "civil-society society" offers other benefits. The value of the concept of a post-capitalist "civil-society society" is less to provide an empirical prediction as social science and more to provide a guidepost and scenario against which progress, or lack of progress, may be measured, and against which rhetorical appropriations of sustainability and social responsibility may be debunked.

If the transition to a "civil-society society" were to occur, it would involve not merely incremental improvements in human rights and remediation of worst practices but instead a fundamental change in the central mission of the publicly traded corporation so that it looks and behaves more like a non-profit organization. As long as the central mission is to maximize short-term returns for an anonymous body of shareholders who in turn buy and sell stocks based on narrowly conceived indicators of financial health, the system is condemned to continue on a treadmill of growth, ecological destruction, and social irresponsibility. Many reforms have been proposed, and resistance to the reforms is often fierce. Governmental policies and civil-society leadership could move the private sector toward practices that are aligned with a vision of just sustainability. However, because the dominant portion of the private sector resists such changes as expensive constraints on a fundamental mission that is premised on short-term earnings, social movements will have to take part in the battles to convert such changes from vision into policy.

Since the 1980s social movements in the United States have lost more political battles than they have won against publicly traded corporations, but social-movement organizations and other civil-society organizations are gradually finding avenues to effect societal change that are directly oriented toward corporate policy. Direct action targeted at publicly traded corporations, such as one finds in IOMs and TPMs, together with indirect action that builds the alternative economics of localist and access institutions, may prove ultimately to be more effective at regulating and changing the corporate world than reform and protest efforts directed at the regulatory policies of the nation-state and global financial institutions, although both together will likely be even more effective. Furthermore, social movements can also contribute directly to the politicization of innovation by continuing to open up to political contestation the priorities of research and development fields, the design decisions for industrial technologies and consumer products, and the contours of technological fields. Likewise, movement-based campaigns can continue to help shift consumer preferences toward products and firms that meet various standards associated with justice and sustainability, such as organic fair-trade products and locally owned import-substituting businesses.

The incipient forms of a shift from production for profit to production to just sustainability standards remain the exceptions in today's economy. The organizational and technological innovations found in the alternative pathways in globalization may be harbingers of a post-capitalist society, or they may be merely small eddies in a tidal wave of ongoing corporate control of every facet of human life. If the former is to become the future, policy changes in international organizations, national governments, private industrial bodies, and local governments are crucial. In turn, the continued mobilization of social movements and other forms of public-interest activism and advocacy will be necessary for creating the political will to make those changes a reality and to stop ongoing attempts at their subversion. Bringing about the changes is likely to require protest action as one aspect of a more comprehensive set of strategies and tactics that also includes scientific research, technological innovation, and entrepreneurship. A transition to a post-capitalist order is not likely to be smooth or pleasant, and it is far from inevitable. The alternative pathways in globalization are neither trivial, romantic experiments nor an exciting harbinger of a postmodern green utopia, but instead laboratories of innovation and experimentation for a contested future. They are incubators of new knowledges and technologies as well as the moral imagination and political will needed for building a just and sustainable society.

Notes

Introduction

1. On the generative dimension of social movements with respect to knowledge, I am developing a perspective found especially in the work of Jamison. See, e.g., Jamison 2001.

2. I view the problem of incorporation and transformation as continuous with the analysis of routinization in Weber's (1978) sociology of religion and the development of the study of routinization for modern political movements by Michels (1958). They are examples of approaches to the historical dialectic that I am seeking to explore in the study of alternative pathways. However, as has been recognized as long ago as the work of Zald and Ash (1966), whose classical analysis of cooptation is also relevant, the iron law of Michels is too restrictive to cover a wide range of historical cases. On incorporation, see also Jamison 2001.

3. The transformations in social movements during the last 30 years of the twentieth century, and their analysis in the "new social movements" literature, can be overemphasized. Identity and lifestyle issues were important in the labor, women's suffrage, abolitionist, and other social movements of the nineteenth century and the early twentieth, and protest has hardly subsided as a repertoire of movement activity after the 1960s. For the statements on new social movements by European theorists, see Habermas 1987, Melucci 1980, 1996, and Touraine 1992. Specific causal explanations or mechanisms that each proposes (respectively, the colonization of the lifeworld, a shift to identity politics, and the end of class politics) capture some of the phenomena. In my mind the deeper contribution of the continental theorists is to draw attention to questions of modernity in ways that are consistent with an historical sociology of modernity. Those questions are sometimes missing in Anglophone social-movement theory, which tends to focus on patterns across time and place. For example, although I see the action/system distinction as characterizing all societal sectors, and I view the types of social movements discussed here as originating in far broader processes than a colonization of the lifeworld, the historical sociological questions that Habermas poses to social-science research are very valuable. For an entry into the literature on shortcomings of at least some versions of new social-movement

theory, see Pichardo 1997; and for an application of new social-movement theory perspectives to issues involving technology and risk similar to issues considered here, see Halfmann 1999.

4. On contentious politics, see McAdam, Tarrow, and Tilly 2001 and Tarrow 1998. As of 2006 there were more than 60,000 websites containing the term "alternative pathways"—with a variety of meanings from the biochemical to the infrastructural—so I claim no originality when I used the term in the web edition of the first part of this book (Hess 2001). At that point Gottlieb 2001 was not available to me, but my use of the term "alternative pathways" is similar to his and attempts a similar comprehensive perspective to the one that he achieved for food and agriculture. Allen et al. (2003) use the related term "alternative food initiatives" and likewise draw on Gottlieb's work; Daley-Harris (2002) also uses the term "pathways"; Feenberg (1995) discusses "alternative modernity"; Henderson (1996) and Pinderhughes (2004) cover a range of alternative pathways in their discussions of "alternative futures"; Luke (1999) discusses "alternative modernities"; Taylor (2000) analyzes four types of pathways in the environmental justice movement; and Alperovitz (2005), L. Brown (2001), Korten (1999), Shuman (2000, 2006), and Williamson, Imbroscio, and Alperovitz (2002) are among the authors whose discussions of alternatives also influenced my thinking. The more entrepreneurial side of the alternative pathways is also approximated by the concept of "building blocks" of the "local living economy" that is found in the networks of the Business Alliance for Local Living Economies (2006a).

5. There are many promising pathways for change, and there are other types of movements that are situated at the intersections of industrial change and the environment. Prominent among them are the blue-green coalitions of labor unions and environmental organizations. Such movements have been analyzed by other researchers and are not considered in detail here. See especially Berg 2003, Gould, Lewis, and Roberts 2004, and Obach 2002.

6. There is a substantial literature on the definitions of social movements and collective action more generally. Melucci's (1996) typology of collective action is well thought out and is constructed in the rationalist, continental tradition that is amenable to a Weberian analysis of ideal typical causal chains. However, the categories of analysis and the distinctive features that I have found useful are somewhat different from his. The specific definition of social movements developed here draws on McAdam and Snow 1997, Snow 2002, and Tarrow 1998. Flacks (2004) emphasizes the distinction between movements and networks of activists.

7. See also definitions that use the phrase "challenge to authority" (Snow 2002), "powerful opponents" (Tarrow 1998), and "articulation of a broader social conflict" (Touraine 1992).

8. On the issue of ambivalence in partnerships between reform-oriented scientists and social-movement organizations, see particularly Clarke 1998, 2000, Frickel 2004a, and Yearley 1992. Allen (2003, 2004) also shows how relations among reform-oriented scientists can exhibit tensions over styles of partnership with activists or social causes.

9. See respectively Weber 1978, Parsons 1977, Durkheim 1964, Marx 1977, Marx and Engels 1973, Wallerstein 1974, and Chase-Dunn and Hall 1997. On the concept of universalism, I follow Parsons's later work (1977: 251), which is more specific than in his earlier work (1951: 82) and which is not cluttered by Weber's expansion of the term "particularism" to cover modern forms of law such as commercial law (e.g., 1978: 698, 880). Weber closely associates universalism with both expansion of scale of a society, such as the Roman empire's "jus gentium" (1978: 696), and with societal differentiation. In other words, the processes are not separate. The Roman law of citizenship was universalistic in a limited sense (Parsons 1977: 94); it encoded hierarchy in a sense similar to the use of the term by Dumont (1977, 1986), even if the law of citizenship contained the kernel of a universalistic, egalitarian, and individualistic legal order. Remnants of the older, hierarchically organized societal categories can be found in modern distinctions between the citizen and noncitizen (Kim 2001). The distinction between social categories that are enfranchised by universalism, in the sense of general rules that apply equally to all parties within a territory, and those that are not is a fundamental point of reference for the legitimacy of access pathways that will be discussed in chapter 6.

10. See especially Beck 1992, 1999, 2000, *Journal of Political Ecology* 2006, and discussions in the anthropology and cultural studies of science on the changing cultural understandings of fundamental biological and natural categories, such as in the work Franklin and Lock (2003) and Haraway (1997). For an amplification of the four theoretical frameworks and an application to the historical sociology of science, see Hess 2006a.

11. On the argument that the various environmental crises that we confront today are global in nature, see Yearley 1996. In the U.S. general public recognition of the environmental crisis is often traced back to the publication of *Silent Spring* (Carson 1962). Diamond's 2005 book *Collapse* provides a good marker of how the general understanding of the societal significance of the environmental crisis developed during the ensuing 50 years.

12. On the democracy deficit, see Held 1996. On the myth of cultural homogenization, see Inda and Rosaldo 2001.

13. Hirst (2000) and Hirst and Thompson (1999) challenge the more extreme statements about globalization while at the same time recognize that significant changes have occurred since World War II. For an analysis of globalization influenced by neoliberal ideology, see Friedman 1999, 2005.

14. On the demise of the nation-state thesis see Castells 1986. Castells endorses it to a degree, whereas Hirst and Thompson (1999) question it.

15. On global cities, see Sassen 2000. On sharing the stage with other actors, see Sassen 2003. Friedman (2003) has often portrayed the change as a dual process of upward and downward integration and fragmentation, and the image is a valuable alternative to simplistic portrayals of the end of the nation-state.

16. On the degree to which the international economy is more internationalized today than before World War I, see Hirst and Thompson 1999. On the growth of financial markets, see Sassen 2003. On the rise and power of global

corporations, see Barnet and Müüller 1974, Barnet and Cavanagh 1994, and Sklair 2001. On international social movements, see Cohen and Rai 2000 and Della Porta and Tarrow 2005.

17. The claim that globalization has increased inequality should be examined comparatively and across classes. Firebaugh (2003) argues that after 1960 (that is, after one date frequently used to mark the beginning of a period described as "globalization") aggregate between-nation inequality decreased whereas within-nation inequality increased (see also the United Nations Development Program reports, e.g., 1999). A more detailed discussion is offered by Schmitt (2000), who argues that almost all of the increase in the inequality of wages in the U.S. between the late 1970s and mid 1990s was due to a decline in real wages at the lower end of the income scale. For example, according to Schmitt women in the top 10 percent experienced gains, but there were greater losses in real wages for women in the lower 10 percent.

18. My interpretation follows Schmitt 2000 and Williamson, Imbroscio, and Alperovitz 2002 by sidestepping the trade-vs.-technology debate and looking instead at the political economy of globalization as a factor. Unionization is one example of lowered wage bargaining power; those statistics are from pp. 37–38 of Williamson, Imbroscio, and Alperovitz 2002. On job losses and the 2004 statistics, see Bronfenbrenner and Luce 2004.

19. On post-Fordism, see Harvey 1989. On university-industry-government partnerships, see Etzkowitz and Leydesdorff 1997. On import substitution, see Shuman 2000, 2006.

20. On diasporization and the failure of the nation-state in its assimilation policies, see Friedman 2003. On ongoing reconstruction of diasporic identities, see Inda and Rosaldo 2001. On ethnogenesis, see Roosens 1989.

21. Weber (1949, chapter 2) provides a succinct statement of the methodology. Ringer's discussion of methodology (2004, chapter 3) is also helpful on this point.

22. See also Fischer's (1995) discussion of types of policy evaluation. A similar division might be constructed for the natural sciences, such as distinctions among natural history, ecology, ecosystem restoration and design, and philosophies of nature.

Chapter 1

1. On the concepts of doxa and fields, see Bourdieu 1971, 1975, 1977, 1982, 1988, 2001 and Swartz 1997. For similar work on conflict in science, see Collins and Restivo 1983. I borrow significantly from Bourdieu, but his lack of a historical perspective on the late-twentieth-century scientific field led to an overemphasis on autonomy, even though he also developed an incipient critique of the autonomy assumption in science studies.

2. I introduced the concept of undone science in a paper (Hess 1998), then explored it in an electronic publication (Hess 2001) and in a co-authored essay

(Woodhouse et al. 2002). See also Hilgartner (2001) on the unknowable; the philosophers of feminist and postcolonial science studies, e.g. Haraway (1989) and Harding (1992, 1993, 1998), who also focus on the systematic exclusion of categories of knowledge and people from the scientific enterprise; and the new political sociology of science (Frickel and Moore 2006; Kleinman 2003).

3. On the metaphor of the left and right hands, see Bourdieu 1998.

4. For an introductory account that includes a discussion of philosophical realism, see Boyd, Gasper, and Trout 1991.

5. This paragraph provides a synopsis of arguments that, for a more leisurely discussion, can be reviewed in Hess 1997 and in such sources as Collins 1983, 1985, Duhem 1982, Kuhn 1970, Lakatos 1978, and Quine 1980.

6. See Collins 1983, 1985 and Kuhn 1970.

7. I am bracketing discussion of his incommensurability thesis, because I assume that the thesis is not borne out by the actual practice of scientists, who are generally able to understand each other's theoretical differences and often willing to negotiate methods. The incommensurability thesis can ground an argument in favor of epistemic relativism, but Kuhn himself rejected that characterization of his work.

8. I cannot review the huge literature here, but my 1997 book *Science Studies*, though now somewhat dated, provides one way into the various currents of the sociology of scientific knowledge. The accounts by Callon (1986, 1994) and Latour (1987) are the best-known and most influential of a large literature on networks, schools, and research programs.

9. On the discussion of core sets and controversy, see Collins 1983, 1985, 2000, 2002.

10. See, for example, Collins's argument about the irreducibility of replication to an algorithm (1985). It is an important argument, but there are points where Collins seems to push it in favor of epistemic relativism, with which I disagree (Hess 1997).

11. I have found metaphors of variation and selection provocative when applied loosely to science and technology. (See, e.g., Basalla 1988, Campbell 1990, and Knorr-Cetina 1981.) However, because there is a danger of over-naturalizing the idea of selection and consequently of blinding the analysis to issues of power, I emphasize an everyday definition of selection. See also Frickel and Moore 2006, Kleinman 2003, and Fuller 2000a. Fuller's discussion of the democratic control of science emphasizes various mechanisms that could enhance democratic control over the selection of research programs.

12. See Merton 1973 (orig. 1957), Hagstrom 1965, and Mauss 1967. On the general consistency of the autonomy assumption with other scientific systems of the modernist period, which tended to emphasize self-correcting and closed systems, see chapter 4 of Hess 1995.

13. See Bourdieu 1975 and Latour and Woolgar 1986. I am retracing somewhat Knorr-Cetina's (1981) discussion of this trajectory of thinking, but with a purpose of leading up to a broader argument.

14. Again, see Latour and Woolgar 1987. Stated more formally, cumulative advantage theory (Merton 1973) holds that students of famous scientists at top institutions start out at the top of the status hierarchies and tend to do better and better over time than students who start out from lower status positions. On actor-network theory, see Callon 1986, 1994 and Latour 1986.

15. For more detailed critiques of Kuhn and the paradigm concept, see Fuller 2000b and Restivo 1983. For medicine the pattern of dominant networks has also been discussed in the analysis of the "dominant epidemiological paradigm" (Brown, Zavestoski, and McCormick 2001; Zavestoski et al. 2002). See also the concept of "scientific and intellectual movements" (Frickel and Gross 2005), which describes a specific type of the more general politics of research fields.

16. On the choices, see Hagstrom 1965: 78–79. On high-risk fields, see Menard 1971 and Hargens and Felmlee 1984. On shifts on interests in the form of band-wagons, see Fujimura 1987. On diminishing returns, see Rescher 1978. On migration and status, see Ben-David and Collins 1966 and Mullins 1972.

17. On priority disputes and assignments, see Merton 1973. On the Matilda effect, see Rossiter 1993.

18. Leading historical studies of the topic include Forman 1987, Kevles 1997, Kleinman 1995, Kohler 1991, Leslie 1993, and Noble 1977.

19. On the quid pro quo, see Clarke 1998, 2000. On relations between scientists and social movements in general, see Frickel 2004a, 2004b, Hård and Jamison 2005, Hess et al. 2007, and Moore 2006a.

20. See my discussion of the phenomenon in a comparison of angiogenesis and cartilage research in the cancer field (Hess 2006a).

21. On invisible colleges, see Price 1963. On the formation of new fields, see Mullins 1972.

22. On diminishing returns, again see Rescher 1978. In the field that has an increase of specialty networks competing over relatively stable funding sources, the specialty networks will engage in various intensification strategies to meet the increased competition for scare resources. Controversy (akin to warfare in societies that face resource shortages) among the specialty networks may break out. We know that controversies occur at the research front, and we know a lot about how they are maintained, negotiated, and resolved, but we know little about the conditions that are likely to produce high and low levels of controversies. The framework outlined here suggests that controversies will be more likely not only at the research front but when competition for scarce resources (funding, journals, etc.) increases.

23. See also Hagstrom 1964, 1965.

24. For a similar argument (the "finalization" thesis), see Böhme, van den Daele, and Krohn 1976 and Schäfer 1983.

25. For more on the critique of the cyclical view of scientific revolutions, see Fuller 2000b.

26. On obliteration and incorporation, see Merton 1973: 508.

Chapter 2

1. On the air pump, see Hessen 1971 and Shapin and Schaffer 1985. On radical groups and their influence on early modern science, see Hård and Jamison 2005 and Jacob 1988. On the alignment of sciences with political and economic agendas, see Forman 1987, Heims 1991, Kevles 1997, Leslie 1993, and Simpson 1998. A related perspective is also being developed as the "new political sociology of science" (Frickel and Moore 2006).

2. On alignment in cancer research, see Fujimura 1987. On the post-Cold War shift in research goals, see Kleinman 1995. On chemistry and early-twentieth-century military and industrial applications, see Kevles 1997 and Noble 1977.

3. On mode-two knowledge production, see Gibbons et al. 1994 and Nowotny, Scott, and Gibbons 2001. On academic capitalism, see Slaughter and Leslie 1997, 2001. On the enterprise university, see Marginson and Considine 2000. On audit culture, see Strathern 2000. On the triple helix, see Etzkowitz and Leydesdorff 1997, 1999 and Etzkowitz, Webster, and Healy 1998. On degrees of compromise, see Croissant and Restivo 2001. On impure cultures, see Kleinman 2003. On the loss of autonomy and the shift in the funding patterns in corporate R&D facilities, see Varma 1995, 1999. In some of the biotechnology firms there may be a trend toward adoption of academic practices and a relative increase in autonomy. The "asymmetrical convergence" thesis of Kleinman and Vallas (2001) notes a collegialization process for small, biotechnology firms that would appear to be different from the trends that Varma found for the large corporate R&D centers. It is but one example of the variation that the study of globalization and science must examine.

4. For a slightly more amplified discussion, see Hess 2001. Education at the K–12 level is also undergoing changes in technology and organization in alignment with the needs of global industry (Monahan 2005).

5. In some cases there are "ET" (environmental technology) goals in strategic plans, but they tend to be configured around patentable and licensable research products, such as fuel cells.

6. On the statistics given, see Engell and Dangerfield 1998.

7. On reflexive modernization, see Beck 1992, 1999, 2000. On ecological modernization, see Mol and Spaargaren 2000. Reflexive modernization is the increased self-consciousness of societies about the hazards and risks of industrial society, and ecological modernization refers to the transformations that are occurring in some industries as they move toward greener production technologies and products. Neither reflexive modernization theory nor ecological modernization theory can be used without awareness of issues raised in the literature. For example, Wynne (1996a) suggests that public risk perceptions are based on judgments of the trustworthiness of expert institutions rather than evaluations of material risk.

8. For a somewhat more detailed discussion of lay authority in the case of alternative medicine, see Hess 2005a. On scientists, activism, and various types of

hybrid organizations, see Frickel 2004b and Moore 1996, 2006a. Brian Martin (e.g., 1996, 2006) has written extensively on whistleblowers and on suppression of dissident scientists; many publications are available at his website.

9. On the exceptionalist policy, see Bimber and Guston 1995. On its historical roots, see Kleinman 1995.

10. I have charted out the two frames in more detail for controversies involving complementary and alternative medicine (Hess 2004a, 2005a).

11. On changes in the theory, methods, and research problem area priorities of primatology due to internationalization and increasing participation by women, see Haraway 1989. On American and Japanese physicists, see Traweek 1988, 1992. On strong objectivity, see Harding 1992.

12. On French and English physics see Duhem 1982. On other examples, see Harding 1998, Harding 1993, and Hess 1995.

13. Farkus (2002) found retrenchment going on during her dissertation field-work of the Dutch science shops during the early 2000s. For a way into the literature of science shops from an American perspective, see Sclove 1995a,b.

14. On community-oriented research and higher education in the U.S., see Strand et al. 2003.

15. For a historical introduction to action research and participatory action research, see Levin 1999 and Puckett and Harkavy 1999. For a discussion of the issue in the context of citizen mobilization and the environment, see Fischer 2000; see also Freire 1986, Greenwood and Levin 1998, Loka Institute 2004a, and the website of PARnet (2006). In the U.S., the Loka Institute developed a network of community-based research, and Cornell University's PARnet provides a gateway to information on the topic.

16. On types of participatory research, especially the continuum between professionally initiated and lay initiated research, see Moore 2006b. On popular epidemiology, see Brown and Mikkelsen 1990. On community-based research and the university see Strand et al. 2003. On the Philadelphia projects, see Puckett and Harkavy 1999.

17. See Agency for Healthcare Research and Quality 2001 and U.S. Environmental Protection Agency 2004b. On participatory research in the public health field, see Green and Mercer 2001 and Israel et al. 1998.

18. On suppression, see again Martin 1996, 1999. On the suppression of high-status scientists, see Richards 1981 and Epstein 1996. In some cases, social-movement mobilization and public pressure from dissident scientists have forced the release of funding for specific areas of undone science, but in such cases the suppression strategy can extend into manipulation of experimental design to introduce biases against the alternative hypothesis or technology (Hess 1999, 2002b; Moss 1996).

19. On the misunderstanding of publics by scientists, see Wynne 1996b.

20. On the differences between survey and ethnographic research, see Wynne 1994. On lay knowledge and reconstruction, see Brown and Mikkelson 1990,

Fischer 2000, Hess 1995, and Irwin and Wynne 1996. On reappropriation, see Eglash et al. 2004. On pockets of literacy and illiteracy, see Hess 1999: 229.

21. On lay involvement in research agenda setting, the key points of entrance into the literature are Brown and Mikkelson 1990 and Epstein 1996. (See also Hess 2005a.) Those examples are from the health field, but similar patterns can be in other fields that are undergoing epistemic modernization, such as in the environmental field (Jamison 2001). More generally, the institutionalization of social movements into interest groups has long been recognized as one general outcome when elites recognize the need for changes that integrate movement demands (Gamson 1990, Zald and Ash 1966). On citizen-scientist alliances, see Brown 2007.

22. On expertification and lay participation, see Epstein 1996 and Hess 1999, 2002b, 2004a, 2005a. For a more general discussion of public participation in technology, see Martin (ed.) 1999. The concept of narrow-band competence (Hess 1999: 229–223) should be viewed as an elaboration of one dimension of the increased lay participation in science; the concept was developed in the context of career shifts that I saw in the movement for complementary and alternative cancer therapies. (See also Hess 2005a.) On falling off the knowledge cliff, see Forsythe 2001. On the recapturing of epistemic challenges by expert groups, see Brown, Zavestoski, and McCormick 2001, Clarke 1998, and Hess 2002b.

23. On demarchy, see Carson and Martin 2002. On citizen panels, see Fischer 2000 and Loka Institute 2004b. One of the American variants involved testing an online consensus conference in comparison with the in-person model (Hamlett 2002).

24. On lay knowledge and its transformations in the CAM cancer therapy movement, see Hess 1999, Hess 2005a, and Hess 2005b. Similar mechanisms can be found in other health social movements, such as the AIDS movement (Epstein 1996) and Gulf War Illness veterans' movement (Brown, Zavestoski, and McCormick 2001). The mechanisms may be generalizable across a wide range of lay groups that mount epistemic challenges.

25. On the use of emeritus professors, see Downey 1988 and Hess 1995: 167–168. On the conversions of some scientists to complementary and alternative medicine research, see Hess 1999 and Hess 2005a.

26. On contributory and interactional expertise, see Collins 2000. On the biographical transformation of researchers, see Hess 1999.

27. In some cases the activist/researcher becomes the lead author or the co-author of a journal publication that advances the research field; the articles can be review essays or presentations of new data. As is common in science, publications are usually co-authored with various experts contributing to different parts of the research project. For examples, see Hess 1999.

28. I use the term "reform movements" because examples of extra-institutional tactics appear to be rare. One confrontation between radical and conservative scientists at a meeting of the American Academy for the Advancement of Sciences became violent when the police were called in to suppress the radicals, but such confrontations are very rare (Beckwith 1986).

29. On scientists, the Vietnam War, and Science for the People, see Beckwith 1986, Krimsky 1982, Moore 1996, Moore 2006a, and Moore and Hala 2002. On later anti-weapons activism, see Gusterson 1996.

30. On the controversy over recombinant DNA, see Krimsky 1982, 1992.

31. Buttel (2006) notes that the failure to generate a research strike, or at least a shift in agendas, involved many factors, including the lack of a political base in small farmers and the general decline in the importance of public, land grant, agricultural research to agribusiness research.

32. On Science for the People, see Beckwith 1986. On the history of the appropriate technology movement, see Kleiman 2000. The development of alternative research fields might compared with "scientific and intellectual movements" (Frickel and Gross 2005) and counter-movements in the sciences (Nowotny and Rose 1979). Alternative pathways in science have similar features to scientific and intellectual movements, but the former may tend to remain more marginalized because they lack high-status leaders.

33. One could also include in this group funding for research on complementary and alternative medicine, which grew dramatically during the 1990s (Hess 2002b, 2004a).

34. On the absolute growth of sustainable agriculture, see Organic Farming Research Foundation 2003. Hassanein (1999) charts the growth of sustainable agriculture research through the 1990s, including the development of research centers, and provides an analysis of an alternative form of research even within this alternative pathway: a more farmer-oriented research model based on farmer networks that occurs in what I call localist pathways in the food and agricultural field. On green chemistry, see Guterman 2000, Woodhouse 2006, and Woodhouse and Breyman 2005.

35. On the suppression of researchers at the National Renewable Research Laboratory, see Carman 2006. However, given the fact that the presidential administration was also courting Midwestern states through the 2006 ethanol initiative, the fired researchers were later rehired. See Holdren 1998, Renner 2001, and Sissine 1999.

36. On the database study and the 0.1 percent figure for funding for organic agricultural research, see Lipson 1997. For a similar study, see Anderson 1995. On acreage, see Organic Farming Research Foundation 2003. On career considerations in the selection of agricultural research fields, see Busch and Lacy 1981 and Goldberg 2001. On the industrial orientation and funding of agricultural research, see Busch 1994. The figure of 1–2 percent of $300 million (for National Science Foundation chemistry research), the absence of research university chemistry departments, and the perception that green chemistry proposals are seen as having a "do not fund" stigma are based on discussions with my colleague Woodhouse, who has researched the topic extensively. See especially Woodhouse 2006.

37. See "Theses on Feuerbach" in Marx and Engels 1973.

Chapter 3

1. On totemism today, see Sahlins 1976 and Hess 1995. On the parallel discussion within technology studies on the politics of artifacts, see Winner 1986. My definitions of technology and design are limited in scope to material culture. People often use the terms metaphorically, for example by referring to policies or administrative practices as "social technologies" or by discussing the design of organizations and policies. However, in order to provide some clarity and limitation of scope, I generally refrain from the metaphorical extensions.

2. On labor-saving devices, and for a critical perspective on how labor-saving technology is linked to increased domestic labor, see Cowen 1983.

3. There are various factors that lead to the "inefficiency" of path dependence, among them the durability of capital equipment, technical inter-relatedness, and the increasing returns patterns of user adoption (Puffert 2003). Some of the exemplars of path dependence, such as the QWERTY keyboard, have been challenged in recent years (Liebowitz and Margolis 1990).

4. On autonomous technology, see Winner 1977, 1986. On gas and electric refrigerators, see Cowen 1999. On closure, see Pinch and Bijker 1999. On sociotechnical systems and momentum, see Hughes 1987.

5. Winner 1977. See also MacKenzie and Wajcman 1999.

6. See Hughes 1983 and Mayntz and Hughes 1988.

7. On the conflict between alternating and direct currents, see Hughes 1983. On actor networks, see Callon 1986. On the concept of a "development arena," which is similar to what I am trying to accomplish with the analysis of technological fields, see Jørgensen and Strunge 2002.

8. See Winner 1986: 80.

9. On the range of definitions of power, see Lukes 1974, 1986. My definition is intended to recognize that power is not always utilized but still exists even when it is not wielded, a point that is not well captured in Weber's classical definition (1978). The definition follows Bourdieu (1977; also Swartz 1997) and Giddens (1995, chapter 6) to the extent that it brings agency and strategy back into the framework, as opposed to Foucault (1980). Actors may be individuals, groups, organizations, or larger aggregates, and the aggregates often include technologies or other non-human entities that have been delegated some degree of agency, as developed in actor-network theory (Callon 1994). However, the delegation of agency to things, such as replacing a police officer with a traffic light, still implies that there are humans with goals somewhere up the chain of delegations that constitute an actor's agency. A definition of power that embraces an expanded concept of the actor is fatally crippled if it does not allow room for the analyst to locate human responsibility and intentionality. The term "resources" encompasses both symbolic and material resources, as well as their interconvertability, as Bourdieu emphasized (1975). The idea of a goal includes the possibility that power can be recursively organized in the ability to shape the goals or agendas of others. (See Lukes 1974, although I part ways with him on the paternalism problem that emerges in his discussion of the third face of power.) Finally, I find

it helpful to categorize goals as values and interests. Interests are the result of an assessment of the effect of an event on the distribution of resources for an actor, whereas values derive from general, legitimating principles about what the ideal outcome of events should be. Although values and interests are usually aligned, they can come into conflict, and actors may sacrifice one for the other, or, for rhetorical purposes, may claim that they are acting from values whereas their opponents are acting from interests.

10. For the classic statement of sociotechnical systems and heterogeneous networks, see Bijker, Hughes, and Pinch 1987. I do not claim great originality in the arguments presented here. For example, the point I make is similar to the arenas analysis of Jørgensen and Strunge (2002), and Nye (1998: 4) notes that existing transportation and energy technologies do not always disappear when new ones are introduced.

11. On privatization of standards, see Reardon and Farina 2002. On the politicization of standards, see Busch 2000.

12. On standards and globalization, see the work of the Institute of Food and Agriculture Standards at Michigan State University, especially Tanaka and Busch 2003. On standards, product design, and safety law, see Jain 2006.

13. See also Hess 2004b,c. Some have asked if the argument is similar to that of the "regulation school" (Boyer 1990), but my argument is much more in a Weberian tradition regarding the growth of universalism under conditions of expansion of scale. The concept of "object conflicts" draws on a research tradition that developed in part from the analysis of "boundary objects" (Star and Greisemer 1989) and "boundary organizations" (Guston 2001). In the context of health social movements, Brown et al. (2004) extend the concepts to point to the role of medical technologies as boundary objects and the role of health social-movement organizations in constructing and maintaining boundary objects across different constituencies. Unlike the boundary objects work, my approach places a greater emphasis on conflict. Closer to my perspective are Clarke and Montini (1993), who show how different social worlds interpret a boundary object differently; Clarke (2000), who shows how the interaction of social movements and maverick scientists leads to product innovation; Winner (1986), who shows how design choices have political implications; and Jørgensen and Karnøe (1995), who show how design choices coincide with differences between social-movement and industrial views of technological and societal development.

14. On regulatory push, see Bayliss, Connell, and Flynn 1998a,b.

15. On the Dutch chemical industry, see Mol 1995. There is a debate over the extent to which the greening of industry has been ecologically significant or dwarfed by the treadmill of growth in aggregate environmental withdrawals and deposits. On ecological modernization, see Mol 1995, 1996, 2000a, 2000b, 2003 and Mol and Spaargaren 2000. On the treadmill of production, see Schnaiberg and Gould 1994 and Pellow, Schnaiberg, and Weinberg 2000. For an assessment and comparison of the two approaches, see Buttel 2000a,b. For a general introduction to the field, see Bell 2004b. A comparative perspective is needed to determine the degree to which specific industries and countries have

undergone ecological modernization or are subject to treadmill-like retrenchment (e.g., Mol and Spaargaren 2000). Freudenberg's (2005) analysis of the disproportional toxic waste generated by a small number of industries (and firms within them) is suggestive of what can be accomplished with industry-specific comparisons. To date, the ecological modernization of industry has not kept pace with increasing levels of consumption and resulting environmental destruction, and consequently social movements have a potentially important historical role to play.

16. On eco-efficiency and eco-innovation, see DeSimone, Popoff, and World Council 1997, Florida 1996, Hawken, Lovins, and Lovins 1999, Porter and van der Linde 1995, and Rennings 2000. Pellow (2002) also develops an analysis of conflicts between the utilization of environmental amelioration frames in the waste industry and the exposure of low-income and ethnic minority groups to industrial hazards.

17. On older struggles, including the issues of politics of design, see Cockburn 1999. On the design of genetically modified food as public biotechnology, see Weeks 2004.

18. One might also use the phrase "alternative technology movements," but the focus on food or financial products makes them as much product-oriented movements as technology movements, and the technologies may sometimes be complementary rather than alternative. I also want to distinguish the term from the "technology movements" of Walsh, Warland, and Smith (1997), whose use of the term is for the environmental justice and anti-toxics movements.

19. See Jamison 2001, Hård and Jamison 2005, and Truffer and Durrenberger 1997.

Chapter 4

1. On polarization, see McAdam, Tarrow, and Tilly 2001. Also see the older literature on cooptation (e.g., Zald and Ash 1966) and on the routinization of charisma (Weber 1978).

2. The urbanization statistics are from U.S. Census Bureau 1995. The other statistics are drawn from calculations by Bonser, McGregor, and Oster (2000: 296–298), which are based on Department of Agriculture data. Although a distinction is often drawn between plough agriculture and horticulture, the terms "agriculture" and "farming" will be used here in a broad sense to include both as well as animal husbandry.

3. See Brieger 2002, Pesticide Action Network North America 2003, and Weir and Schapiro 1981.

4. On the anti-GM-food campaigns in general, see Reisner 2001. On the California protests, see Weintraub and Gogoi 2003. At the time of writing, the history of anti-GM-food activism was in better condition for Europe (see, e.g., Purdue 2000) than for the U.S. A good archive of news articles on the topic is provided by GMWatch (2006). On the Kentucky mobilizations, see Fitzgerald 2005.

5. On Monsanto and Greenpeace, see Greenpeace 2004.

6. On the decision to ban a category of antibiotics, see Union of Concerned Scientists 2005. On opposition to factory farms in general, see Citizens Environmental Coalition and Sierra Club 2005, Dobb 2000, and Grace Factory Farm Project 2006. On the suppression of researchers, see Brady 2003 and Winne 2006.

7. On the Community Environmental Legal Defense Fund, see Linzey and Grossman 2004.

8. On anti-GM-food campaigns' focus on corporate targets, see Schurman 2004.

9. On national identity and anti-GM-food mobilizations in Europe, see Harper 2004.

10. On Detroit, Gofman, and the Union of Concerned Scientists, see Joppke 1993: 27–29. On the Sierra Club, Friends of the Earth, and Committee for Nuclear Responsibility, see Wellock 1998: 31–105. Gofman later co-authored an influential book (published by the Sierra Club) on the risks of diagnostic and therapeutic radiation in medicine (Gofman and O'Connor 1985).

11. On new orders, see Moyer et al. 2001: 140. On Consolidated National Intervenors and Mothers for Peace, see Joppke 1993: 30–32, 87. On the Creative Initiative Foundation, see Wellock 1998: 157–161.

12. On Seabrook and direct action, see Joppke 1993: 81–87 and Moyer et al. 2001: 143.

13. On the Construction Work in Progress laws and shift in orders, see Moyer 2001: 142–146. On costs, see Joppke 1993: 135. Campbell (1988: 6–9) analyzes the variety of reasons for the demise of the industry: inept management and design, increased government safety regulation, decreased demand due to the effects of OPEC on prices, and (his own contribution) a decentralized state that was accessible to anti-nuclear groups. The mood of the movement changed rapidly in the late 1970s. I remember joining protests at Diablo Canyon and Rancho Seco, and by that time the protests were less political and more educational and festive, with interest already being shown in nuclear weapons.

14. On Brown and Carter, see Joppke 1993: 68–70, 139. On Sundesert, see Wellock 1998: 173–177. On the 1980s, see Joppke 1993: 149–157.

15. On relations between anti-nuclear-energy and anti-nuclear-weapons organizations, see Joppke 1993: 145–148. On the beginnings of the anti-nuclear-weapons movement, see Gusterson 1996. On science and politics regarding nuclear winter, see Martin 1988.

16. See Friends of the Earth 2006, Natural Resources Defense Council 1998, Public Citizen 2006a,b, Sierra Club 2006a, and Union of Concerned Scientists 2006. Power plants, incinerators, and transportation are the major generators of air pollution in the U.S., but their relative contributions vary from city to city. At an aggregate level power plants and transportation generate 80 percent of carbon monoxide (mostly from transportation), 72 percent of carbon dioxide (about evenly split), 75 percent of sulfur dioxide (about evenly split), and 75 percent of nitrogen oxides (mostly power plants); see Creech and Brown 2000:

197–199. On emergent divisions within the environmental movement over nuclear energy, see Little 2005.

17. U.S. Department of Energy 2003, 2006.

18. On the history of American environmentalism during the 1960s and the 1970s, see Dowie 1995, Gottlieb 1993, and Kline 2000. On air pollution, see Dupuis 2004. Gottlieb draws attention to the earlier history of activism around occupational health and the precursors of what would be considered today to be the environmental justice movement.

19. I am following Gottlieb (1993) and extrapolating somewhat on his analysis.

20. See Dowie 1995 and Gottlieb 1993.

21. On the Environmental Defense Fund and emissions trading, see Dowie 1995. On eras in American environmental policy, see Mazmanian and Kraft 1999. The Group of Ten included the Environmental Defense Fund, the Environmental Policy Institute, Friends of the Earth, the Izaak Walton League, the National Audubon Society, the National Parks and Conservation Association, the Natural Resources Defense Council, the National Wildlife Federation, the Sierra Club, and the Wilderness Society. In 2005 I surveyed the websites of those organizations to see how much they were focused on the issue of industrial pollution control. The preservationist organizations were less concerned with issues of pollution and toxics, but some of them had programs on clean water, clean air, and/or pesticides and toxics, usually with a focus on wilderness issues such as acid rain, wilderness water quality, pollution in parks, and pesticides in birds. In contrast, there was greater concern with issues of industrial pollution and environmental change in the Environmental Defense Fund, Friends of the Earth, the National Resources Defense Fund, and the Sierra Club. The interest was also prominent in organizations that were outside the Group of Ten but played a significant role in second wave of the environmental movement, such as Greenpeace, Public Citizen, and the Union of Concerned Scientists. As with some of the "Group of 10" organizations, Greenpeace also diversified from marine preservationism and anti-nuclear politics. In the early 1990s Greenpeace worked with environmental organizations around the Great Lakes to push the International Joint Commission to recommend sunsetting chlorine-based chemicals. Industry rallied with a massive campaign on both sides of the Great Lakes that framed the issue as a jobs-versus-environment tradeoff, and by 1995 the issue had been buried in both Washington and Ottawa. See J. Howard 2004 and Thornton 2000.

22. On the various strands or streams that have fed into the environmental justice movement in the U.S., see Cole and Foster 2000 and Pellow and Brulle 2005.

23. On the distinction in the social address of the environmental justice movement and the anti-toxics movement, see Brulle and Pellow 2006. On the growth of environmental justice organizations, see Dowie 1995: 133. On the organizational diversity of environmental justice organizations, see Brulle and Essoka 2005. On the anti-toxics movement and Superfund legislation, see Szasz 1994. Cole and Foster (2000) do not include the emergent environmental health movement in their survey of streams, but it is clearly connected with many environmental justice issues. See Brown et al. 2002, Gibbs 2002, and McCally 2002.

24. On the chronology of environmental justice events in the 1990s, see Newton 1996, Brulle and Pellow 2006, and Summit II National Office 2002b. On the opening of political opportunities during the Clinton administration, see Taylor 2000.

25. On the report, see the U.S. Environmental Protection Agency 2004a.

26. On PIBBY, see Bullard 1994. On NIABY, see Heiman 1990 and Dowie 1995: 133.

27. On the second summit and the broadening of goals, see Summit II National Office 2002a.

28. See Shellenberger and Nordhous 2004 and Brulle and Jenkins 2006.

29. On the claim that air quality has declined in some parts of the U.S. since the 1970s, and on the role of fossil fuels in air pollution, see Natural Resources Defense Council 2003. Industrial air pollution is a major source of pollution in some places, such as Louisiana's "Cancer Alley" (Allen 2003; Roberts and Toffolon-Weiss 2001), but in many cities power plants, incinerators, and transportation are the main sources. On the concentration of pollution in hot spots, see Dowie 1995.

30. On the conditions for success based on case studies in Louisiana's "Cancer Alley," see Roberts and Toffolon-Weiss 2001. See also Walsh, Warland, and Smith 1997.

31. On the shift from environmental justice to sustainability goals, see Agyeman 2005a,b, Alternatives for Community and Environment 2005, and Arbor Hill Environmental Justice Corporation 2006. Pellow (2002, 2005) examines the shift that some recycling advocates underwent from pro-incinerator to anti-incinerator stances and explores some of the connections between environmental justice and sustainability issues such as extended producer responsibility.

32. On the greening of bus emissions, see Bus Riders Union 2004ab, Hess and Winner 2005, and Hess 2006b. Not all cities followed the pattern that I am describing; in Los Angeles the environmental justice/transit activists have maintained the transition to CNG.

33. On the development of sprawl, see Kunstler 1996 and Hott 1997. On urban growth machines and coalitions, see chapter 3 of Logan and Molatch 1987. The designs of American metropolitan spaces, and their attendant technologies, can be contrasted with the more compact urbanism of Europe (Beatley 2000).

34. On San Francisco, Miami, and Baltimore, see Mohl 2004. On San Francisco, see also Issel 1999. On Overtown, see Hott 1997.

35. On Boston, see Lupo, Colcord, and Fowler 1971. On the effectiveness of the coalition in San Francisco, see Mohl 2004.

36. On the relationship between anti-highway mobilization and calls for increased funding for public transportation during the 1960s and the early 1970s, see Lupo, Colcord, and Fowler (1971).

37. For an example of suburban anti-highway campaigns, see Sunnocks 2001. On Sprawl-Busters, see Norman 2006a,b. On the airport campaigns, see Regional Commission on Airport Affairs 2005. The Sierra Club (2006b) main-

tains links to some of the organizations and anti-sprawl campaigns. I do not discuss the infrastructure politics of rural areas, such as hydroelectric facilities, which in the U.S. were often situated on lands occupied by Native Americans.

38. The already large literature on the anti-corporate, anti-globalization movement is growing rapidly. For a diagnosis of the issue of corporate control, see Barnet and Müüller 1974 (the classic statement and still a fine introduction), Barnet and Cavanagh 1994 (a follow-up that contains a good discussion of international banks), Danaher and Mark 2003 (which has a chapter on the early- and mid-twentieth-century history of challenges to corporate power in the U.S., Korten 1995 (which analyzes a range of related issues), and Perkins 2004 (which reveals the close linkages between foreign loans and foreign policy).

39. On the shift in protest from the South to the North, see Podobnik 2004 and Smith 2004. On the long-term historical roots of the movement opposing corporate globalization, see Broad and Heckscher 2003 and Podobnik and Reifer 2004. On the trajectory of the movement, see Buttel and Gould 2004.

40. The protest against the Chico dam project in the Philippines in the mid 1970s (Broad and Heckscher 2003) may have been the first major anti-dam protest mobilization against the World Bank. Other protests followed with respect to development projects in China, India, and Brazil. A particularly dramatic case was the Polonoreste project in Brazil, which helped to build roads that opened up a huge area of the Amazon (Price 1989). The attention drawn to Polonoreste, such as Price's exposé (which he told me circulated in World Bank circles and had some influence on their policies), increased pressure on the World Bank to reform its policies. In addition to the many good sources in Portuguese and Spanish on the Amazonian basin, see Cockburn and Hecht 1989. On the responses of the World Bank to environmentalist pressure, see O'Brien et al. 2000: 122–133. On the sense that activists had overestimated the World Bank's influence, see Clarke 2003: 105.

41. On the contrast with the World Bank and the failure of the WTO, see O'Brien et al. 2000: 120, 141–153. On environmentalists' participation in Seattle and the emergence of blue-green coalitions, see Berg 2003, Gould, Lewis, and Roberts 2004, Obach 2002, and Rose 2000. On NAFTA's notorious chapter 13 provision and its effects on national sovereignty and national and state/provincial environmental legislation, see Moyers 2002. The Methanex case, which Moyers discusses, was eventually rejected on a technicality. Regarding environmental issues, one of the primary Southern organizations has been Vía Campesina, a farmer-peasant coalition that has drawn attention to the negative effects of WTO policies on local environments and incomes and has called for removing agriculture and food from the WTO altogether. (See Edelman 2005.)

42. Monroe Friedman, the leading scholar of boycotts in the U.S., found 24 ecological boycotts between 1987 and 1992, the period of time studied in his book (Friedman 1999). Half of the boycotts involved animal rights and half involved environmental protection. In contrast with other types of boycotts, Friedman found that ecological boycotts tend to target producers, not retailers; have a national scope and long duration; and be media oriented. The smaller groups of environmental boycotts (the Waste Oil Action boycotts directed at Chrysler and

General Motors for recycled oil) are not covered here. See also O'Rourke 2005. (O'Rourke studied campaigns directed at Staples, Nike, and Dell.)

43. See Danaher and Mark 2003 and Earth Island Institute 2004.

44. Rainforest Action Network 1996, 1998, 2002, 2006b.

45. Rainforest Action Network 2006a.

46. Gerefi, Garcia-Johnson, and Sasser 2001.

47. Another type of direct action aimed at corporate investment policies that can involve environmental issues is shareholder activism (increasingly termed "advocacy"), but here we do not see the same dynamic in the incorporation and transformation process. The first American shareholder resolutions around justice and environmental issues were moved in 1969 and 1970. A group of physicians moved a resolution against Dow Chemical's production of napalm in Vietnam (Welsh 1988), and Ralph Nader led a resolution in favor of corporate responsibility at General Motors. In 1971 the Episcopal Church presented General Motors with a resolution to withdraw from South Africa, and other resolutions on South Africa, infant formula, and tobacco soon followed. By the 1990s the meager beginnings had mushroomed into hundreds of shareholder resolutions. One of the factors behind the growth of shareholder resolutions was the increasing percentage of equity held by institutional investors. Mutual funds, foundations, and pension funds tended to join together in backing shareholder resolutions, often on corporate governance issues, and religious and environmental organizations, together with socially responsible mutual funds, were behind the increases in resolutions on social and environmental issues (Gillan and Starks 2000, 2003, Robinson 2002). By the early 2000s there were 75–85 shareholder resolutions on environmental issues alone (Thomsen 2001). Although the resolutions often failed, they served as a useful tactic in broader campaigns to change corporate behavior, and in some cases they were withdrawn when the corporation agreed to negotiate a solution.

48. For more on the weakness of the locally oriented strategy for oppositional movements, see Gould, Schnaiberg, and Weinberg 1996.

Chapter 5

1. For more examples, see my discussions of the TPMs around complementary and alternative medicine, the Danish wind movement and industry, and the free-libre open-source movement (Hess 2005b).

2. Jamison (2001) has noted the process of "incorporation" of environmental social-movement goals into business practices. Hård and Jamison (2005) also discuss the general process of appropriation of scientific research and technological innovation. My analysis is similar to theirs and builds on their analysis by drawing attention to the concomitant transformations of the design of technologies, products, and technological systems.

3. The Community Alliance with Family Farmers, founded in late 1970s as the California Agrarian Action Project, began by supporting farm workers who had

been put out of work due to the mechanical tomato harvester. In subsequent decades the organization developed campaigns for pesticide legislation; opposition to genetically modified food; conferences and other work around organic and ecological farming, including support for research on sustainable agriculture; and labels and campaigns in support of local food (Allen et al. 2003, Community Alliance with Family Farmers 2003, Guthman 2004).

4. On the origins of organic agriculture in the 1920s and on its development in the United Kingdom, see Conford 2000. On the relationship between organic farming and industrial agriculture, see Buck et al. 1997, Guthman 2004, Hess 2004c, and Kaltoft 2001. On the early history of the movement, see Conford 2000. On the claim about the first modern use of the term "organic" in Northbourne 1940, see Lotter 2000. Guthman (2004: 4–6) also points to sources for the organic foods movement in the permanent agriculture movement that arose out of the Dust Bowl and the concerns with food purity that were especially prominent in the first 30 years of the twentieth century, as well as more recent influence from the environmental and "back to the land" movements.

5. The history of the mid-twentieth-century American organic movement is still incomplete, and it may turn out that the focus on Rodale that I have taken is misplaced. Biographical material on Rodale is available from Greene (1971) and Jackson (1974), and other background is available from the Rodale Institute (2006). Peters (1979) provides a scholarly account, and Lotter (2000) and Conford (2000) also provide some information. Conford (2002) also suggests that Rodale's response from the scientific community may not have been as hostile as portrayed by Peters (1979). My claim that Rodale's original vision of the organic was relatively technical and depoliticized is from reading his 1948 book *The Organic Front* and the portrayals of him by Greene (1971) and Jackson (1974). It would likely be revealing to read through issues of his magazine from that period. In contrast, Northbourne's 1940 book sounds more of the themes that would become prominent in the late-twentieth-century local agricultural networks. See also Hadwinger (1993) and Peters (1979) on the interest of Secretary of Agriculture Henry Wallace in using manure and organics in the soil. The sources differ somewhat regarding the name of the original magazine, its changes of title over the years, and the details of the mailings and the response by farmers.

6. Although the organization was a firm and eventually became a successful publishing company, during the early years the magazine was supported by Rodale's electrical wiring company, and he also started a foundation in 1947. The ambiguity of classification is an example of the mixture of advocacy goals and profit-oriented production that is characteristic of TPMs. On the finances of Rodale during the early period and the role of his son Robert, see Greene 1971, Haberen 1997, Jackson 1974, and Peters 1979. In Europe during the 1940s several advocacy organizations were also founded, including the Soil Association, and some segments of the British organic movement had fascist leanings (Conford 2000, Reed 2001). The leanings appear to be absent in the American movement, probably because of the New Deal connections of Wallace and Rodale's upbringing as a Jew (he changed his name from Cohen).

7. See Northeast Organic Farming Association of Vermont 2006 and Guthman 2004: 112–113.

8. On the missions of the organic associations, see California Certified Organic Farmers 2003, Northeast Organic Farming Association of New York 2006, and Northeast Organic Farming Association of Vermont 2006. On the history of the appropriate technology movement, see Kleiman 2000. For more on the confluence of dietary therapies in the medical field with the sustainable agriculture movement, see Hess 2002a.

9. See chapters 2–5 of Guthman 2004.

10. On the organizational conflicts and developments, see Organic Trade Association 2006 and Guthman 2004: 114–115, 134. Although organic cotton farming was still a very small niche, a group of organic cotton farmers formed the Texas Organic Cotton Marketing Cooperative (Cordes 2003). The coop generated its own business ventures, and it attracted an organic textile mill that relocated from Vermont. From a broader perspective the Organic Trade Association estimated that there were only about 43 organic cotton farmers in the U.S. in 2004, and acreage had decreased after the implementation of the federal organic program in 2002. In addition to difficulties encountered with the organic standards, farmers pointed to foreign competition as a reason for decreasing acreage. Many of the farmers sell in foreign markets, and they must compete with foreign suppliers in both domestic and foreign markets (Organic Trade Association 2004, 2005). However, by 2006 the situation was changing due to the nonprofit organization Organic Exchange, which was helping to develop organic cotton markets.

11. For a journalistic account of the trends, see Pollan 2001. On the changes in agricultural technologies, see Guthman 2000. The industrialization of organic agriculture has been fairly well studied in the literature (e.g., Allen and Kovach 2000, D. Goodman 2000, Goodman and Goodman 2001, Guthman 1998, Klonsky 2000). *Nutrition Business Journal* (e.g., 2001, 2004) tracks many of the trends from the food perspective, but as an industry journal it is expensive and hard to access. Dupuis (2000, 2002) has covered the dairy industry and health concerns with bovine growth hormones.

12. On the Cascadian Farms story, see Pollan 2001. On industrial acquisitions, see P. Howard 2006.

13. For details on the Stonyfield case in the context of dilemmas experienced by small, progressive businesses, as well as the Ben and Jerry's experience, see Hollender and Fenichell 2004: 236–237. Ben and Jerry's had a more localist and justice orientation and a less environmental orientation (it was mostly non-organic until the acquisition), so the value clash was predictably stronger in their case than with Stonyfield. For related comparisons of clashes between profitability and social responsibility goals, see Weinberg 1998. On the decline of price premia that has accompanied consolidation, see Smith and Marsden 2003.

14. On consolidation of the conventional food sector, see Schwartz 2005. On the figure of 73 percent for conventional sales, see Schneider 2005. On the growth of the food, drugs, and mass channel, see Spencer and Rea 2003. On Wal-Mart, see Gunther 2006. On commodity chain analysis of organic foods, see Buck, Getz, and Guthman 1997.

15. On sustainable local agriculture as a social movement, see Hassanein 1999. On the bifurcation thesis, see Campbell and Liepins 2001 and Guthman 2002.

16. On various categories of organic production and the complexities of defining what is organic, see Guthman 1998.

17. My analysis of the politics of functional foods is based in part on attendance at the Tenth Annual Conference of the Functional Foods for Health Program, University of Illinois at Chicago Circle, June 2001.

18. On the politics of food definitions, see also Goodman and Dupuis 2002. Lockie, Lyons, and Lawrence (2000) and DeSoucey (2004) discuss broader categories of food and their framing, and DeSoucey in particular draws attention to movement/market tensions in food politics. DuPuis (2000) discusses the intersections of health concerns and food categories in a study of bovine growth hormone and milk. Guthman (2004: 122) also notes that the NutriClean program provides a pesticide-free alternative to the organic label.

19. On the organic standards controversy, see Vos 2000 and Organic Consumers Association 2006. On the harmonization process, see EnviroWindows 2002. On the synthetic ingredients provision, see *Consumer Reports* 2006.

20. For more on the conflicts around the definitions of the term "organic," see D. Goodman 2000.

21. On the solar entrepreneurs and energy industry, see Reece 1979.

22. See Hayes 1979, Pollack 1984, and Reece 1979.

23. I am relying on Allen 2000 for the historical information, and also on my own memories of working to publicize a solar energy tax credit program in California during this period. Gorman and Mehalik (2002) describe some of the problems with early solar hot water heating systems. From 1995 to 2001, sales of solar energy equipment grew sixfold to $2.5 billion (Griscom 2001).

24. On windmills in California and on windmill factories, see Asmus 2001: 29–31. In Denmark mobilizations against nuclear energy also spilled over into pro-wind activism, but wind energy also had a longer history of activism that can be traced back to progressive rural movements in the early twentieth century (Jamison et al. 1990, Jørgensen and Karnøe 1995). I discuss the Danish wind energy movement as a TPM, with a focus on the incorporation and transformation process, in Hess 2005b. Likewise, in Germany opposition to the Whyl nuclear plant led to the formation of a solar-oriented energy policy around Freiburg, Germany, which has since become a world center for solar research and the solar industry (Solar Region Freiburg 2003).

25. On hippies, see Gipe 1995: 90. On sobriquets and references to Governor Jerry Brown, see Asmus 2001: 137.

26. On wind-energy advocacy by environmental organizations, see Mayer, Blank, and Swezey 1999.

27. On wind-energy funding preferences and Heronemus, see Asmus 2001. On the general history of the wind industry, see Asmus 2001 and Righter 1996.

28. On the Danish industry, see Jørgensen and Karnøe 1996 and Jørgensen and Strunge 2002. On problems of wind energy, see Righter 1996.

29. On recent developments, see Gipe 2002.

30. On the gap in wind-turbine production, see Cooperman 2003. On weatherization, see National Center for Appropriate Technology 2006.

31. On universities and investment in clean technology, see Blumenstyk 2003. On the hydrogen drain on renewable-energy research, see Mieszkowski 2004.

32. See National Grid 2005, which I selected as an example because it is the provider that services the area where I live. Note that some states set a limit on the scale of hydropower capacity when including it in renewable portfolio standards. On participation rates in green pricing programs, see U.S. Department of Energy 2004.

33. On problems faced by small wind-energy providers, see Cooperman 2003.

34. On environmentalists' opposition to wind energy, see Asmus 2001: 138–140.

35. For critiques of the recycling movement, see Luke 1997 and Weinberg, Pellow, and Schnaiberg 2000.

36. On early-twentieth-century recycling practices, see Hickman 1999, Melosi 1981, Scheinberg 2003, and Strasser 1999. Seldman (1995) describes the economic basis of the shift away from locally oriented recycling during the twentieth century.

37. For the historical background, see Seldman 1995 and Lounsbury, Ventresca, and Hirsch 2003.

38. On the partnership with a non-local and increasingly transnational remanufacturing industry, see Gould, Schnaiberg, and Weinberg 1996. On the exposure of low-income workers to workplace hazards, see Pellow 2002. Note that the arguments here may be specific to the U.S. and other wealthy countries. Grassroots recycling operations appear to be growing in other countries that do not have the same political economy of waste and may have different implications regarding job quality (e.g., International Council for Local Economic Initiatives, n.d., Case Studies 3, 21, and 60).

39. On the statistics for incinerators and on the Reagan incentives, see Blumberg and Gottlieb 1989: 34–80. For the general history, see Seldman 1995 and Lounsbury, Ventresca, and Hirsch 2003. On the Keep America Beautiful campaign and its industrial backing, see Weinberg, Pellow, and Schnaiberg 2000: 16–17. On Westinghouse and other major firms involved in nuclear energy reactors, see Walsh, Warland, and Smith 1997: 4 and Blumberg and Gottlieb 1989: 51.

40. On the different organizations and the rise and fall of incineration, see Lounsbury, Ventresca, and Hirsch 2003. On local recycling organizations and incineration, see Pellow 2002 and Walsh, Warland, and Smith 1997. On the integrated strategy as a way to mute opposition, see Blumberg and Gottlieb 1989: 191.

41. See Lounsbury, Ventresca, and Hirsch 2003.

42. On recycling statistics, see R. W. Beck, Inc. 2001. On the marginalization of community-oriented recycling programs, see Weinberg, Pellow, and Schnaiberg 2000.

43. On workplace hazards and labor issues, see Pellow 2002 and Weinberg, Pellow, and Schnaiberg 2000. On the Californian and Canadian recovery rates, see Connett and Sheehan 2001. On downcycling and the increase in absolute generation of solid waste, see Spiegelman 2003. On the comparison with Japan and Denmark, see Seldman 1995.

44. See Connett and Sheehan 2001 and Motavelli 2001.

45. Among the North American cities, counties, and provinces that had adopted zero waste policies by 2003 were Del Norte and Santa Cruz Counties, California; Seattle, Washington; Toronto, Ontario; and the province of Nova Scotia (Connett and Sheehan 2001, Motavelli 2001). On European and Japanese take-back and extended producer responsibility legislation, see Palmer and Walls 2002.

46. On the computer take-back programs, see Computer TakeBack 2005 and Grassroots Recycling Network 2001, 2002. On related environmental justice issues in Silicon Valley and computer take-back campaigns, see Pellow and Park 2002 and Pellow 2005. On the zero emissions program, see Rainforest Action Network 2006c.

47. My comments on sustainability in industrial design in the U.S. are anecdotal, but they are based on interactions with industrial and other product design ers through a product design program that I have taught in. Their assessment on the lack of priority of sustainability issues, especially in the U.S., in contrast with Europe, is based on conferences, journals, and knowledge of design firms. On the potentials and challenges of green chemistry as an alternative research field and industrial technology, see Woodhouse 2005 and Woodhouse and Breyman 2005. On the risks of nanotechnology, see Etc. Group 2004a,b.

48. On the various government procurement standards, see Government Purchasing Project 2006.

49. On extended producer responsibility versus product stewardship, see Palmer and Walls 2002. On the company that changed its tune on take-back after exposure of toxic waste dumping in China, see Hollender and Fenichell 2004: 137.

50. On the relationship with progressive reform movements and the role of women reformers, see Wirka 1996. On the Regional Planning Association of America and its relationship with other planning traditions, see Talen 2005. On the post-World War II reformers, see Gans 1959, Jacobs 1961, and Mumford 1961, 1963.

51. See Hoffman 1989 and Pyatok 2000. The Planners Network, founded in 1975, had dwindled by 1980.

52. The intellectual roots of New Urbanism have antecedents in various strands of twentieth-century planning history (Stephenson 2002, Talen 2005). For the charter and a brief history, see Congress for the New Urbanism 2006a,b.

53. The politics of the new urbanists are far from straightforward, and discussions sometimes distinguish between a more conservative, neotraditional East Coast strand and a more environmentally oriented West Coast strand (O'Keefe 2002). Andres Duany and Elizabeth Plater-Zyberk of Seaside fame are representative of the former; the 1986 book by Sim Van der Ryn and CNU founding

member Peter Calthorpe, *Sustainable Communities*, is an example of the Western strand. That book explicitly developed an ecological and social critique of suburban sprawl that showed some continuity with the concerns of the grassroots housing, anti-sprawl, and urban environmental movements. On urban growth boundaries and housing costs, see Pozdena 2002. On New Urbanism as a reaction to policy over form, see Fulton 1996 and M. G. Brown 2002.

54. For a review and a visual presentation of the showcase projects, see Katz 1993.

55. On the growth of neotraditional developments and the price premia, see Kadet 2005. On the Gaithersburg, Maryland, case, see Whorisky 2005.

56. For the sobriquet "new suburbanism," see Harvey 2000: 170. On critiques and appraisals of new urbanism, see Krieger 1998, Pyatok 2000, and the broader discussion, mostly positive, in Bressi 2002. On the defense of new urbanism's relevance to urban poverty issues, see Bohl 2000.

57. See Elliott, Gotham, and Milligan 2004.

58. On planners originally concerned with sprawl and density in new urbanism, see Krieger 1998. On the regional level, see Calthorpe and Fulton 2001 and Marshall 2000.

59. On the New Urbanism Division, see American Planning Association 2005. On the confluence of the two, see Zimmerman 2001. On smart growth in general, see Gillham 2002, Knapp and Talen 2005, Smart Growth Network 2003, and Smart Growth America 2006. On urban growth machines, see Logan and Molatch 1987.

60. One indication of the broadening of the agenda was that in 1989 the American Institute of Architects Committee on Energy was renamed the Committee on the Environment. In 1992 the committee published the *Environmental Resource Guide*, and in the following year the professional organization's conference theme was sustainability. In 1993 the newly formed United States Green Building Council held its first conference, which took place in conjunction with that of the architects. The Green Building Council brought together the chair of the Committee on the Environment with representatives of the construction and home appliance industry, as well as the Rocky Mountain Institute. Likewise, during the mid 1990s the Clinton administration and various federal agencies supported building analyses and greening. See Cassidy 2003a,b and American Institute of Architects, Committee on the Environment 1992.

61. On LEED standards and statistics, see Cassidy 2003a and U.S. Green Building Council 2005.

62. On the industry code, see National Association of Home Builders 2005. On the variation among residential codes, see Moore and Engstrom 2005. On the systemic approach, see H. Levin 2005.

63. On the position of the research fields, I took a "key informant" approach and relied on the insights of experienced colleagues in the fields of planning and architecture. As in the case of my suggestions regarding product design and industrial design in the previous section, the observations put forward here can

only be considered hypotheses. According to one colleague in planning (a senior person in a leadership position), planners originally rejected new urbanism, but later embraced it even as they tended to downplay recognition for the Congress of New Urbanism; the ambivalent relationship may be due to professional rivalries between planners and architects. Within the field of planning, concern with "smart growth" is quite mainstream and on the agenda of researchers in the best planning departments. In contrast, according to a colleague in architecture (a journal editor with substantial knowledge of the field, including sustainable design), in the U.S. the high-status private architecture schools still tend to focus on deconstruction and postmodernism, whereas interest in sustainability is more prominent in the public universities, including some of the high-status ones.

64. On the politics of zoning standards, see Kunstler 1996.

65. On the Puritan and Pax Funds, see Hollender and Fenichell 2004. On the Interfaith Council on Corporate Responsibility, see Robinson 2002 and Wolf 2004.

66. Investment screening refers to criteria that institutions or individuals use to eliminate some companies from their portfolios. For background information and the statistics cited, see Social Investment Forum (2001). The Social Investment Forum—the national trade organization for socially responsible investing in the U.S.—defines the field of socially responsible investing to include investment screening, community investing, and shareholder activism. Although socially responsible mutual funds may engage in shareholder activism (they prefer the term "advocacy"), I would classify the type of action under the oppositional movements, because such work does not generate an alternative product per se but instead calls for a change in corporate policy, usually in the form of an end to some kind of production practice or investment priority.

67. For a review of socially responsible mutual funds, see Social Investment Forum 2005. On the Calvert-Vanguard partnership, see Johansson 1999 and Dow Jones Sustainability Indexes 2005.

68. See Social Investment Forum 2001.

69. See Grimes 2000, 2005, Raynolds 2000, 2002, and Transfair USA 2006.

70. See Grimes 2000, 2005, Daviron and Ponte 2005, Raynolds 2000, 2002, and Transfair USA 2006.

71. See Daviron and Ponte 2005 and Ponte 2004.

72. On the success of eco-labels in Europe, see Boström 2006.

Chapter 6

1. The literature is extensive. The following sources are particularly relevant: Robertson 1995 (on glocalization), Castells and Hall 1994 (on the technopole), Etzkowitz and Leydesdorff 1997, 1999 (on the triple helix), Sassen 2000, 2003 (on the global city), and Saxenian 1996 (on Route 128 and Silicon Valley). For more on the comparison of green localism and the green technopole, see Hess 2004b and Winner and Hess 2007.

2. Alperovitz 2005, Shuman 2000, 2006. For an earlier formulation of the import-substitution strategy at a metropolitan level, see Jacobs 1969.

3. The term "local" is left open to a variety of levels below that of the large nation-state, from neighborhoods to small states within a federal system, such as the state of Vermont, or small nation-states, such as Denmark, within the European Union. In practice, I often use the term "local" to refer to a city or metropolitan region that encompasses the equivalent of several counties in the American political system and may include millions of inhabitants. Shuman (2006) suggests a definition based on the smallest unit of tax authority, and there are many benefits for using his definition in a policy-making context. However, I also want to leave open the ability to consider small-scale projects in domestic units and at the neighborhood level, such as ecovillages.

4. On poor people's movements, see Piven and Cloward 1971, 1977.

5. On the features of localism, see Shuman 2000, 2006. Just as social movements for access to the basic material stuff of life have a long historical legacy in American political culture, so the concept of economic localism in the U.S. could be traced back to Jeffersonian democracy and even to the models of self-sufficiency found in the subsistence economies of the Native American peoples and the colonial and frontier settlements. In the twentieth century, Mumford (1961) drew attention to the possibilities of a human scale of urban infrastructure, and Schumacher (1973) developed the tradition during the era of the counterculture and the appropriate technology movement.

6. Activist intellectuals such as Morris (2001), Shuman (2000), and Alperovitz (Williamson, Imbroscio, and Alperovitz 2002) developed the concept of local economic sovereignty as a way of moving beyond left and right political agendas while also providing an alternative to dependence on publicly held corporations as an economic base for a region. Shuman (2006) especially has developed the economic rationale for the import-substitution strategy.

7. The pattern of transition from advocacy work to service provisioning is recognized in the literature on access action in specific fields, such as Morgan 2002 (for health) and Poppendieck 1998 (for food), but I believe this chapter is the first comparative, cross-field analysis of the pattern. On incorporation and transformation of poor people's movements, see Piven and Cloward 1977. See also the discussion of access organizations in health social movements by Brown et al. (2004).

8. See Bonnano et al. 1994 and Bonser, McGregor, and Oster 2000.

9. See Aley 2004, American Agriculture Movement 2001, Browne 1983, and Ritscher n.d.

10. There is a substantial network of social-science researchers who have studied the topic in some detail. This section builds on their work and examines it from a different framework that emphasizes innovation and historical change. See especially Allen et al. 2003, Bell 2004a, D. Goodman 2000, Goodman and Dupuis 2000, Guthman 1998, 2000, 2002, and Hassanein 1999.

11. The statistics in the paragraph and the shift to intensive agriculture are from Heimlich and Anderson 2001.

12. For examples of the non-profit urban farm, see the Portland and Sacramento case studies in Hess and Winner 2005. On community and conservation land trusts, see Witt and Rossier 2000.

13. In a CSA, consumers become members of a farm and buy a subscription for a season. Shares are usually dropped off at designated spots on a weekly basis. Consequently, consumers eat what is ripe and harvested for the week, share in the risks of farming, and benefit from good seasons, just as they receive less food or less variety during bad seasons. The claim that CSAs originated in Japan is apparently erroneous, according to McFadden (2004a), who interviewed the founders of the first two farms. Anthroposophy is a spiritual philosophy developed by Rudolf Steiner and was mentioned in chapter 5 as one of the roots of organic farming. On the estimate of up to 1,700 farms, see McFadden 2004a,b. (The number may be high.) On the role of women in CSAs, see Cone and Myhre 2000. On the size of CSA farms and farm income, see, e.g., Stevenson et al. 2004. On the aggregation of subscriptions, see Galayda 2006.

14. Although the chain stores had taken a significant portion of the market for natural foods, the total food sales volume for natural products retailers in 2002 was $10.4 billion (with about 44 percent of it organic), which means that a significant portion was still in independent stores that were neither cooperatives nor chains. The historical material and statistics are from National Cooperative Grocers Association 2002, Spencer and Rhea 2003, and Swanson, Nolan, and Gutknecht 2001.

15. On definitions and types, the discussion of the 1976 legislation, and statistics from 1970 to the late 1990s, see A. Brown 2001, 2002. On the 2000 study, see U.S. Department of Agriculture 2000. On the sales statistics for 2000, see Bullock 2000. On growth statistics for 2004, see U.S. Department of Agriculture 2004. On farmers' markets as incubators, see Feenstra et al. 2003 and Economics Institute 1999. Growth has also been dramatic in other countries. For example, in the United Kingdom the number of farmers' markets went from none in 1997 to 300 in 2001 (BBC News 2001).

16. On the Local Hero campaign, see Community Involved in Sustaining Agriculture 2005. On the German label, see chapter 3 of Brand 2006. Other "buy local" campaigns are documented in Hess and Winner 2005. They are examples of a broader phenomenon in which consumption has become politicized around sustainability and local ownership issues (Cohen 2005, Cohen and Murphy 2001, Princen, Maniates, and Conca 2002, Shuman 2006, Spaargaren and Van Vlit 2000).

17. The percentage of meals eaten away from home increased from 16 in 1977–78 to 29 in 1995 (Lin, Frazao, and Guthrie 1999).

18. See Green Restaurant Association 2002a,b and Chefs Collaborative 2005. Guthman (2002) also discusses the role of locally oriented restaurants, including Chez Panisse.

19. On the Organic Valley cooperative, see Strohm 2003 and Organic Valley Family of Farms 2006. On Appalachian Sustainable Development, see Flaccavento 2002. On Cascadian Farms, see Pollan 2001. On the transformation

of the Mondragon system, see Huet 2001. On consolidation more generally, see Howard 2006.

20. The discussion of the national anti-hunger organizations is based on chapter 7 of Eisinger 1998. Some of the organizations receive funding from corporate donations (generally those oriented toward charitable distribution of food), and some corporate foundations also address hunger issues.

21. On the history of anti-hunger organizations and movements in general, see Poppendieck 1998. On Reagan-era policies such as the Temporary Emergency Food Assistance Program, see Gottlieb 2001: 208. On food banks, see Cotugna and Beebe 2002 and Jacob 2003.

22. On the critique of the change of frameworks from food provisioning as a "right" provided by food stamps to food provisioning as a charitable activity, see Poppendieck 1998. On community food security, see Allen 2004, Gottlieb 2001, and Gottlieb and Fisher 1996. On dependency relations, see Curtis 1997, Poppendieck 1998, and Tarasuk and Eakin 2003.

23. Community gardening is a subset of the broader international trend toward urban agriculture. Bryld (2003) notes that the urbanization process is only one of the variables shaping the growth of the urban agriculture movement. The negative effect of structural adjustment programs and the lack of government funding for urban poverty programs have also contributed to near-starvation levels of poverty in some urban centers of the poorer countries (Rosset 2001). Despite assassinations and other forms of repression, land reform has been carried out successfully in some countries, and it is gaining increasing recognition, even if in a modified form, among national governments (Langevin and Rosset 1997, Rosset 2001). An exception to the general lack of government support for urban agriculture has been the Cuban government (Murphy 1999, Warwick 2001). On the complex relationships with wage labor and gender, see Bryld 2003, Murphy 1999, and United Nations Development Program 1996.

24. On the history of community gardens I have relied on Von Hassell (2002), who cites R. Goodman (2000) on the World War II period and Bassett (1979) on the post-1960s period. Another good general source for the history is Lawson (2005).

25. Again, see Von Hassell (2002: 142), citing R. Goodman (2000) and the American Community Gardening Association (1998). Lawson's statistics (2005: 241), which are based on two surveys by the American Community Gardening Association in 1990 and 1996, are somewhat different: 2,123 new gardens were created during that six-year period (a 35 percent increase), but 542 were lost. The figure of 6,000 gardens is from Lawson (2005: 241). In the United Kingdom the number of community gardens had grown to more than 500 by the end of the 1990s (Guardian 2000). For our project's case studies of community gardens, see Hess and Winner 2005. Although the numbers may seem significant, they are relatively small in comparison with urban agriculture in developing countries. For example, the total number of community gardens in the U.S. may be less than that just one Latin American city, Havana, where gardeners have benefited from much stronger support from the national government.

26. I am summarizing results from our case studies (Hess and Winner 2005) as well as published research and reports by Von Hassell (2002), Lawson (2005), and Madsen (2002).

27. On the Vermont local food franchise, see Shorto 2004.

28. For case studies of the public power agencies of Seattle, Sacramento, and Austin, see Hess and Winner 2005.

29. I am going partly on memory from working for Kucinich in a political campaign during the 1970s, but see also Kucinich 2004. On San Francisco, see Hess and Winner 2005. There is also a great deal of information on the website of the American Public Power Association (2006) as well as on the websites of specific publicly owned municipal utilities.

30. On local power in general and the California developments, see Fenn 2004 and Hess and Winner 2005.

31. On the "back to the land" movement, see Jacob 1997: 3, 53. On the technological dimension of the movement and its relationship to the development of renewable energy, see also Kleiman 2000. On the home-power movement generally, especially its non-economic dimension, see Tatum 1994, 1995, 2000. Although wood and biodiesel may be technically classified as renewable because the plants eventually grow back and recapture carbon, both generate greenhouse gases. Even if one accepts the argument that they are carbon neutral in the long term, they are not scalable as a general solution to home heating needs.

32. The comments are based on attendance at Solarfest in Vermont on July 14, 2001. The total exchange systems of archaic markets and town fairs (Mauss 1967) bear comparisons with farmers' markets, solar festivals, community garage sales, flea markets, time banks, and other types of local economic exchange institutions (see Hermann 1997).

33. The magazine *Mother Earth News* also has significant coverage of off-grid, renewable energy projects, and the company Real Goods provided an important supply source and catalog. On the growth in sales of solar equipment and statistics for the magazine readership and solar equipment, see Griscom 2001. Actual sales figures for the magazine were one-half to one-third the readership level, but they were still growing during the early 2000s (Home Power 2006a). On the increases in wind energy, see Gipe 2002. On British Petroleum Solar in the home-power market, see Griscom 2001 and Tatum 2000.

34. On guerilla solar, see the columns in the magazine *Home Power* and its website (Home Power 2006b). On grid sellback and net metering, see Cooperman 2003. On certification of installers, see Tatum 2000: 126.

35. On LIHEAP, see Gish 2001 and Kaiser and Pulsipher 2002. On the organizations mentioned, see National Fuel Funds Network 2006b and National Center for Appropriate Technology 2003. On direct assistance by public power utilities, see Hess and Winner 2005.

36. See National Fuel Funds Network 2006a and National Low Income Energy Consortium 2006.

37. The federal government provides energy assistance to low-income people through two major programs: the Weatherization Assistance Program and

LIHEAP. On low-income weatherization, see National Center for Appropriate Technology 2003. On energy and low-income families, see Creech and Brown 2000: 192–193. On green affordable housing, see Green Affordable Housing Coalition 2006. On low-income solar energy programs, see U.S. Department of Health and Human Services 1999. I have also found two examples of low-income green pricing programs from electric utilities: the Los Angeles Department of Water and Power's program "Green Power for a Green L.A.," which claims that 50 percent of its 80,000 users are low-income (U.S. Department of Energy 2004), and the Clean Energy Choice program of the Massachusetts Technology Collaborative (2006). A related phenomenon is the NativeEnergy green tags program, which many progressive firms (or subsidiaries) have joined (see Co-op America 2003 and NativeEnergy 2006). Green tags are a form of renewable energy currency, which can also be purchased for donation to non-profit organizations. Energy conservation utilities, which are funded by a charge on the consumer utility bill, also have a dedicated portion of expenditures for low-income customers.

38. On the concern with deregulation and public power, see the case study of Seattle City Light in Hess and Winner 2005.

39. On the lack of portability of the Danish model, see Desrochers 2001. On stalled projects, see Portney 2002. On regional eco-industrial networks, see Schlarb 2001. On the trend toward science parks, see National Alliance of Clean Energy Business Incubators 2002. For an overview of some projects to date, see Chernow 2002. Arguably the closest to a large eco-industrial park in the U.S. is the Intervale in Burlington, Vermont, but when I visited the project it was still based primarily on composting. There was talk of expanding it to use waste heat from the local wood-burning electricity-generating plant.

40. Lach (2000) provides the American statistics cited and adds that the social address of the yard sale economy is still unknown, but we know that the older age brackets (59 percent for 45–54-year-olds) buy more than average, and young families both sell more (24 percent) and buy more (50 percent) at yard and rummage sales. For Hermann's estimate of the market size and the phenomenon of resale on the Internet, see Efrati 2006.

41. On urban government concerns about tax-free businesses, see Maher 2003. The New York example refers to the town of Ballston Spa, based on personal observation, and the phenomenon is also occurring in other area cities. The city of Sunnyvale, California, sponsors garage sales as a strategy to help reach the state's requirement of 50 percent waste diversion (Russell 2000). Collective yard sales are also an occasion for strengthening civil society ties (Herrmann 1997). In the United Kingdom there is a parallel phenomenon of "car boot" (car trunk) sales, where people drive cars to a designated location and sell to buyers from their cars. They grew rapidly during the 1990s (Gregson and Crewe 2003) but may have subsided in the subsequent decade.

42. On the transformation of recycling into reuse and the Chicago center, see Weinberg, Pellow, and Schnaiberg 2000. On Construction Junction, see the case study by Rachel Dowty in Hess and Winner 2005. I am pulling a few highlights

from the case studies that we developed and the various interviews that we conducted in 2005. Some reuse stores also sell remanufactured furniture.

43. See Green Institute 2006 and the case studies of reuse centers in Hess and Winner 2005.

44. On the Salvation Army, see Horne and Maddrell 2002 and Warren 1999.

45. On the trends in the United Kingdom, see Horne and Maddrell 2002. On the Value Village/Savers chain, see Seavy 2000. The sector's growth in the late 1990s and the early 2000s had slowed from the 1980s and the early 1990s. The Salvation Army (2002) estimates the total number of its thrift stores internationally to be about 1,400.

46. See National Association of Resale and Thrift Shops 2006. One example of a consignment shop chain is Snooty Fox, which sells fashionable clothing and which grew at a rate of 13 percent per year in the late 1990s (Miller 1999). There is also a network of approximately 50 "furniture banks"—organizations that pick up furniture and household items and then distribute them back to people who cannot afford furniture (Furniture Bank 2006).

47. In my research for this book and other projects, I visited some flea markets, including one in San Jose that claims to be the largest in the U.S. The market had a largely Latino clientele, and the stalls looked very similar to the street vendor fairs that one can find in Latin America. Most vendors were selling inexpensive, imported, new goods rather than used goods. In contrast, some towns in the region where I live have supported the growth of antiques districts as a revitalization strategy.

48. See the case studies in Hess and Winner 2005.

49. See the Chattanooga case study in Hess and Winner 2005. On biofuels, localism, and ownership issues, see Morris 2006 and Institute for Local Self-Reliance 2006.

50. Again, see the Chattanooga case study in Hess and Winner 2005, Morris 2006, and Institute for Self-Reliance 2006.

51. On car-free cities, see Crawford 2000 and Worldcarfree.net 2006. On the claim for the superior efficiency of bicycles, see Wilson 1973. The history of pedestrian and bicycle activism is not yet developed, but there are some good beginnings (Batterbury 2003, Blickstein and Hanson 2001, Demereth and Levinger 2003, Levinger 2002).

52. On straw-bale construction, see Henderson 2003, 2006. On the LEED standards, see U.S. Green Building Council 2006.

53. Cohousing facilities generally involve closely situated row houses or apartments in larger buildings, with streets replaced by pedestrian walkways. There is a common area where meals, recreation, and other activities can be shared, and members of a cohousing unit commit to sharing at least some meals on a weekly basis, even though they have kitchens in their own homes. See McCamant and Durrett with Hertzman 1994.

54. On difficulties of bank financing and extensions of the model to lower income groups, see McCamant, Durrett, and Hertzman 1994: 235–236. On the growth statistics, see Cohousing Association of America 2003.

55. The professional histories of cohousing and ecovillages have yet to be written, so the literature is much less developed than, say, the literature on sustainable local agricultural networks. On the history, see Global Ecovillage Network 2005b, Jackson 2004, and Jackson and Jackson 2004. On ecovillages in the U.S., see Ecovillage Network of the Americas 2006.

56. On the definition and goals of the movement, see Global Ecovillage Network 2005a, Jackson 2004, and Jackson and Svensson n.d. On participation in anti-globalization meetings, see Jackson and Jackson 2004. On urban projects, see Urban Ecovillage Network 2004 and EcoCity Cleveland 2004.

57. On the statistics, which are from 1990, see Bullard, Johnson, and Torres 2000a. On transit racism in general, see Bullard, Johnson, and Torres 2000b, 2004.

58. The rail-bus tradeoff is particularly intense in Los Angeles. See Berkowitz 2005 and Bus Riders Union 2004a,b.

59. See Hess and Winner 2005 and Hess 2006b.

60. See Dreier 1997, McNeely 1999, O'Connor 1999, Rusk 1999, Stoutland 1999, and Weir 1999.

61. See Dreier 1997, McNeely 1999, O'Connor 1999, Rusk 1999, Stoutland 1999, and Weir 1999. According to Dreier (1997), the leading organizations in the housing movement in the 1990s were the National Coalition for the Homeless, the National Low-Income Housing Coalition, the National Community Reinvestment Coalition, and the National Congress for Community Economic Development. Other important organizations included ACORN, National People's Action, and the Industrial Areas Foundation. Bockmeyer (2003) also describes the decline of federal funding during the 1980s and the 1990s and its effects on CDCs and housing activism. Stoecker (1997; cf. Bratt 1997) describes Robert Kennedy's early vision and some of the financial constraints that led many CDCs to shift their missions.

62. On the changes in CDC goals, see McNeely 1999, Stoutland 1999, Weir 1999, and Williamson, Imbroscio, and Alperovitz 2002. On the lack of influence on the general trend toward concentration of poverty, see Rusk 1999. On the changes in the late 1990s, see Stoutland 1999 and Williamson, Imbroscio, and Alperovitz 2002: 213–223. The statistics, from National Congress for Community Economic Development 2001, are based on a census in 1998. For a sense of the scope of CDCs, the 500,000 homes built by CDCs compares with 150,000 homes built by Habitat for Humanity worldwide by 2004 (Habitat for Humanity International 2006b).

63. Here I am summarizing arguments presented in Bockmeyer 2003.

64. Cooperative housing involves residential ownership of units, and in affordable cooperative housing there are restrictions on how much profit individuals can earn on resale. Unlike cohousing there is usually limited communal activity such as meal sharing. On the history of affordable cooperatives, I rely mainly on Sazama 2000. On the National Association of Housing Cooperatives, see also Siegler and Levy n.d. On the statistic of 1.5 million cooperative units, see the National Association of Housing Cooperatives 2006.

65. On green affordable housing initiatives, see Global Green USA 2006, Green Affordable Housing Coalition 2004, and Habitat for Humanity 2006a.

66. For a commercial bank with a mandate for "environmentally sustainable community development," see Shorebank Pacific 2004.

67. For the aggregate statistics, see National Credit Union Administration 2006a,b and World Council of Credit Unions 2004. In low-income countries, low dividend rates (sometimes below local inflation rates) and lack of ability to withdraw funds resulted in poor track records for savings. Consequently, many credit unions did not generate sufficient funds from savings to meet the demand for loans. Instead, in many countries during the 1970s and the 1980s, credit unions relied on international donors, which had selected credit unions as vehicles for channeling funds to farmers and small businesses. As a result, credit unions became dependent on donor agencies and passed along credit to borrowers whose potential for default was high. When donors shifted to other institutions as channels for aid programs, many credit unions suffered a liquidity crisis. See Magill 1994: 146, 149 and Lennon and Richardson 2002: 92–93.

68. On consolidation, see Wilcox 2005. On New York, see Schlett 2006.

69. See Wysocki 2006 and Credit Union National Association 2003.

70. The typology is based on talks given at the 2004 conference on "Local Currencies in the Twenty-First Century," hosted by the E. F. Schumacher Society. The literature on local and complementary currencies is extensive; Greco 2001 is a good introduction. For an ethnographic account of Ithaca Hours, see Maurer 2005.

71. The observations are based on notes from the 2004 conference on "Local Currencies in the Twenty-First Century."

72. See Mitchell 2001, 2003, American Independent Business Alliance 2004, and Boulder Independent Business Alliance 2004. On the fear of losing customers, see Hess and Winner 2005.

73. See Business Alliance for Local Living Economies 2006a,b and Hess and Winner 2005. A parallel endeavor is the emergence of local green business directories (Co-Op America 2006), but they are not restricted to locally owned, privately held businesses.

74. The National Community Capital Association (2005) has a good comparison of the five types of community development financial institutions.

75. On the history, see National Federation of Community Development Credit Unions 2006b.

76. See National Federation of Community Development Credit Unions 2006a,b. In 2001 the total assets in the U.S. in community development financial institutions (that is CDCUs, community loan funds, community venture-capital funds, and community banks) were $7.6 billion. The figures and their growth rates—the aggregate figure tripled from 1999 to 2001—may seem impressive, but the total is less than 1 percent of the total socially responsible investments. On the aggregate figures, see Social Investment Forum 2001: 20–23.

77. See Greco 2001 and Time Dollar USA 2006a,b. I have also relied on notes from a lecture by Edgar Cahn (2004).

78. See North 2003, Seyfang 2003, 2004, and Time Dollar USA 2006a.

79. On the Grameen model, see Berenbach and Guzman 1994 and Yunus 1999.

80. For the statistics, see Microcredit Summit Campaign 2006, White and Campion 2002, and the dust jacket on Yunnus 1999. On the commercialization process more generally, see Christen and Drake 2002 and Rhyne 2002.

81. On the connections with neoliberalism, presidential administrations, and legislative initiatives, see Jurik 2005: 8, 49–70 and Carr and Tong 2002. On the role of gender, see Jurik 2005: 70 and Hung 2002. On the trade association and statistics for growth, see Jurik 2005: 64–70.

82. On the differences with low-income countries and problems encountered in the U.S., see Carr and Tong 2002, Jurik 2005: 72–77 and Sevron 2002. On welfare-related problems in Illinois, see Yunus 1994: 176, 185. On the 1996 legislation and the effects of work requirements, see Sevron 2002.

83. On conflicts and trends within microfinance, see Carr and Tong 2002, Sevron 2002, and chapters 5 and 6 of Jurik 2000.

84. Shuman (2006) is developing guidelines for a localist approach to spending preferences.

85. The discussion of BALLE is based on notes from the 2005 annual meeting in Vancouver, on an interview with Shaffer (the "Local Exchange" case study in Hess and Winner 2005), and on Shuman 2006.

86. On the response of the state to poor people's movements, see Piven and Cloward 1971, 1977. On devolution and privatization, see Bockmeyer 2003, Campbell 2000, Dreir 1997, Jurik 2005, Morgan 2002, and Poppendieck 1998.

87. On informal knowledge networks in the agricultural field, see Bell 2004a and Hassenein 1999. In making the claim that there is little formal scientific research in support of localist innovation, two qualifications are necessary. First, as my citations in this chapter demonstrate, social scientists have produced research on the economic and social dimensions of localist pathways, such as rural sociologists for local agricultural networks. Second, in some cases the natural science, engineering, and design research fields associated with the TPMs provide research that is valuable to the localist pathways, as in the case of organic agricultural research. However, in other cases the design of technologies is substantially different in the TPM and localist pathways, such as grid-scale wind turbines versus home-power microturbines, or New Urbanist neighborhood plans versus ecovillage plans.

Conclusion

1. In putting forward this argument, I am building on work on social movements, science, and technology by Brown and Zavestoski (2004), Epstein (1996), Fischer (2000), Frickel (2004b), Jamison (2001), Martin (1999), Moore (2006a), Winner (1986), and Woodhouse and Breyman (2005).

2. See Marx and Engels 1973, Schumpeter 1975, Giugni 1998, and Giugni, McAdam, and Tilly 1999.

3. For classic references on the issue of routinization and cooptation, see Weber 1978, Michels 1958, and Zald and Ash 1966. See also Pynchon 1973. On the distributive/redistributive dimension as it is applied to social movements, see Schnaiberg 1982, 1983a. Schnaiberg argued that although the appropriate technology movement had a populist rhetoric that emphasized the democratic potential of technological innovations (such as solar energy), in practice the movement was more concerned with the distributive politics of shifting resources into soft technology rather than the redistributive politics of enhancing equity. (See also Schnaiberg 1983b, 1983c; Winner 1986.) Schnaiberg's use of the distributive/redistributive distinction is based on but not identical to the discussion by Lowi (1964, 1972).

4. On "greenwashing," see Beder 1997.

5. The definition of sustainability in the "Brundtland Report" (World Commission on Environment and Development 1987) only draws attention to justice in the sense of resource access across generations, but other sections of the report discuss the potential and need for a broader and global understanding of justice. On the "just sustainability" approach, see Agyeman, Bullard, and Evans 2003 and Agyeman 2005b. Agyeman and his colleagues examine linkages between sustainability and environmental justice and review the literature on the topic. Also influential for me are O'Connor's (1998) discussion of the red-green relationship, McGranahan and Satterthwaite's (2000) discussion of the parallel brown-green relationship, and, in the planning context, Campbell 1996 and Moore 2007. My approach is parallel to that of Campbell (1996) in that he also draws attention to the relationship between justice and environmental sustainability, but I differ in assuming that the third E (efficiency or economy) is a means to the end. Because the justice component requires an economy to satisfy basic human needs, a strictly economic or efficiency criterion is taken here to be a side effect or implication of just sustainability, rather than a separate or independent dimension to sustainability. In other words, an economy must be able to provide jobs and meet the standards that it sets for human rights, but economic viability is a means to an end, not an end in itself. See also Sen's (1999) capabilities-oriented concept of development.

6. On zero waste, see Pauli 1998. On industrial ecology, see Chernow 2002. On biomimicry, see Benyus 1997. On cradle-to-cradle, see McDonough and Braungart 2002. On closed-loop manufacturing, see Hawken, Lovins, and Lovins 1999. On living machines, see Todd and Todd 1993.

7. Since the middle of the twentieth century, non-profit organizations have undergone explosive growth. Along with the expansion of the sector has come differentiation in function, so that some non-profit organizations have developed missions that approximate the value of just sustainability. Drucker (1993: 175–176) claims that the non-profit and voluntary sector is the largest employer in the U.S., but the claim does not mean that non-profit employment is the only or primary employment for most persons. Salamon and Sokolowski (2004) estimate the size of the non-profit sector in 1995 was about 6.3 percent of aggregate primary employment (but with an additional 5.2 percent of volunteers) and 7.5 percent of GDP. For a deeper understanding, those numbers must be viewed alongside the growth trends. Non-profit organizations have grown from about

13,000 in the U.S. in 1940 to more than 1.5 million during the next 50 years, and internationally from about 1,500 in 1950 to more than 25,000 in 2001 (Van Till 2000: 74–75, citing Hall 1992: 62 for the 1940 statistic). By the end of the twentieth century the non-profit sector in the U.S. was estimated to be as large as one third of the for-profit sector in terms of number of organizations and about 5–10 percent of the economy in terms of revenue and employment (Bennett and DiLorenzo 1989: 16–17, Weisbrod 1988: 168).

8. Shifts in sectoral dominance should be analyzed with the understanding that one should avoid templating all societies into a unilineal evolutionary sequence, as was characteristic of nineteenth-century anthropology and has been widely rejected in the twentieth century. The critique of the shortcomings of unilineal sequence models has, for cultural and social anthropology, made any kind of general historical analysis or development of typologies a rather marginalized venture, and consequently the work is left to archaeologists and sociologists. (See, e.g., Chase-Dunn and Hall 1997.) On the history of the publicly traded corporation, see Micklethwait and Woodridge 2003.

9. For the influences on this discussion, see Beck's (1992, 1999) thesis of "reflexive modernization" and Evans's (2002) discussion of convergence of civil society and the state. See also Kleinman and Vallas (2001) on the "asymmetric convergence" of the private sector and the academy. I am also drawing on Mol's argument that ecological modernization will require the emancipation of environmental values from economic ones, as well as the arguments of treadmill of production theorists who question the extent to which such emancipation has occurred or can occur within the constraints of the publicly traded corporation's profit mission. See Mol 1995, 1996, 2000a, 2000b, 2003, Pellow, Schnaiberg, and Weinberg 2000, Weinberg 1998, and Weinberg, Pellow, and Schnaiberg 2000.

10. For some sample discussions of the controversy over shifting role of civil-society organizations and their move into service provision and competition with for-profits, see Clarke 2003, Bennett and DiLorenzo 1989, and Weisbrod 1998.

11. On the option of going private to achieve social and environmental responsibility goals, see the case of Seventh Generation and the discussion by Hollender and Fenichell (2004). For discussions of post-corporate economies, see Alperovitz 2005, Gunn 2003, Korten 1999, Shuman 2006, and Williamson, Imbroscio, and Alperovitz 2002. The emancipation of values that Mol (1995, 2000b) discusses is more evident in the non-profit sector, and the possibilities are also being expressed, in incipient forms, in the corporate-responsibility movement and the discussion of a transition toward the "civil corporation" (Zadek 2001). Booth (1998) also argues that the economic structure of producer cooperatives tends to encourage lower resource consumption than privately held firms.

References

Agency for Healthcare Research and Quality. 2001. "Community-Based Participatory Research: Conference Summary." Retrieved March 31, 2006 from http://www.ahrq.gov.

Agyeman, Julian. 2005a. "Alternatives for Community and Environment: Where Justice and Sustainability Meet." *Environment* 47 (6): 10–23.

Agyeman, Julian. 2005b. *Sustainable Communities and the Challenge of Environmental Justice.* New York University Press.

Agyeman, Julian, Robert Bullard, and Bob Evans. 2003. *Just Sustainabilities: Development in an Unequal World.* MIT Press.

Aley, Ginette. 2004. "American Agriculture Movement." In *Encyclopedia of American Social Movements*, ed. Immanuel Ness. Sharpe Reference.

Allen, Arthur. 2000. "Prodigal Sun." *Mother Jones*, March-April: 64–69.

Allen, Barbara. 2003. *Uneasy Alchemy: Citizens and Experts in Louisiana's Chemical Corridor Disputes.* MIT Press.

Allen, Barbara. 2004. "Shifting Boundary Work: Issues and Tensions in Environmental Health Science in the Case of Grand Bois, Louisiana." *Science as Culture* 13 (4): 429–448.

Allen, Patricia. 2004. *Together at the Table: Sustainability and Sustenance in the American Agrifood System.* Pennsylvania State University Press.

Allen, Patricia, and M. Kovach. 2000. "The Capitalist Composition of Organic." *Agriculture and Human Values* 17 (3): 221–232.

Allen, Patricia, Margaret Fitzsimmons, Michael Goodman, and Keith Warner. 2003. "Shifting Plates in the Agrifood Landscape: The Tectonics of Alternative Food Initiatives in California." *Journal of Rural Studies* 19 (1): 61–75.

Alperovitz, Gar. 2005. *America Beyond Capitalism: Rebuilding Our Wealth, Our Liberty, and Our Democracy.* Wiley.

Alternatives for Community and Environment. 2005. "About ACE." Retrieved March 31, 2006 from http://www.ace-ej.org.

American Agriculture Movement. 2001. "History of the American Agriculture Movement." Retrieved March 31, 2006 from http://www.aaminc.org.

American Community Gardening Association. 1998. *National Community Gardening Survey*. City of New York, Department of General Services, Operation Green Thumb.

American Independent Business Alliance. 2004. "AMIBA History." Retrieved March 31, 2006 from http://amiba.net.

American Institute of Architects, Committee on the Environment. 1993. *Environmental Resource Guide: Subscription*. American Institute of Architects.

American Planning Association. 2005. "New Urbanism, Renewed Neighborhoods: Our History." Retrieved March 31, 2006 from http://www.planning.org.

American Public Power Association. 2006. "American Public Power Association." Retrieved March 31, 2006 from http://www.appa.org.

America's Second Harvest. 2004. "Food Banking." Retrieved March 31, 2006 from http://www.secondharvest.org.

Anderson, Molly. 1995. "The Life Cycle of Alternative Agricultural Research." *American Journal of Alternative Agriculture* 10 (1): 3–9.

Arbor Hill Environmental Justice Corporation. 2006. "Arbor Hill Environmental Justice Corporation." Retrieved March 31, 2006 from http://www.ahej.org.

Asmus, Peter. 2001. *Reaping the Wind: How Mechanical Wizards, Visionaries, and Profiteers Helped Shape Our Energy Future*. Island.

Associated Press. 2003. "Richest Got Richer between '92, 2000." Retrieved March 31, 2006 from http://www.priorities.org.

Barnet, Richard, and John Cavanagh. 1994. *Global Dreams: Imperial Corporations and the New World Order*. Simon and Schuster.

Barnet, Richard, and Ronald Müller. 1974. *Global Reach: The Power of Multinational Corporations*. Simon and Schuster/Touchstone.

Basalla, George. 1988. *The Evolution of Technology*. Cambridge University Press.

Bassett, T. J. 1979. Vacant Land Cultivation: Community Gardening in America, 1893–1978. M.A. thesis, University of California, Berkeley.

Batterbury, Simon. 2003. "Environmental Activism and Social Networks: Campaigning for Bicycles and Alternative Transport in West London." *Annals of the American Academy of Political and Social Science* 590 (1): 150–169.

Bayliss, Robert, Lianne Connell, and Andrew Flynn. 1998a. "Sector Variation and Ecological Modernization: Towards an Analysis at the Level of the Firm." *Business Strategy and the Environment* 7 (3): 150–161.

Bayliss, Robert, Lianne Connell, and Andrew Flynn. 1998b. "Company Size, Environmental Regulation, and Ecological Modernization: Further Analysis at the Level of the Firm." *Business Strategy and the Environment* 7 (5): 285–296.

BBC News. 2001. "Marketing the Markets." September 7. Retrieved March 31, 2006 from http://news.bbc.co.uk.

Beatley, Timothy. 2000. *Green Urbanism: Learning from European Cities*. Island.

Beck, Ulrich. 1992. *The Risk Society: Towards a New Modernity*. Sage.

Beck, Ulrich. 1999. *World Risk Society*. Polity.

Beck, Ulrich. 2000. *What Is Globalization?* Polity.

Beckwith, Jon. 1986. "The Radical Science Movement in the United States." *Monthly Review* 38 (1): 118–128.

Beder, Sharon. 1997. *Global Spin: The Corporate Assault on Environmentalism*. Chelsea Green.

Bell, Michael. 2004a. *Farming for Us All: Practical Agriculture and the Cultivation of Sustainability*. Pennsylvania State University Press.

Bell, Michael. 2004b. *An Invitation to Environmental Sociology*. Pine Forge.

Ben-David, Joseph, and Randall Collins. 1966. "Social Factors in the Origins of a New Science: The Case of Psychology." *American Sociological Review* 31 (4): 451–465.

Bennett, James, and Thomas DiLorenzo. 1989. *Unfair Competition: The Profits of Nonprofits*. Hamilton.

Benyus, Janine. 1997. *Biomimicry: Innovation Inspired by Nature*. William Morrow.

Berenbach, Shari,and Diego Guzman. 1994. "The Solidarity Group Experience Worldwide." In *The New World of Microenterprise Finance*, ed. María Otero and Elisabeth Rhyne. Kumarian.

Berg, John, ed. 2003. *Teamsters and Turtles? U.S. Progressive Political Movements in the Twenty-First Century*. Rowman and Littlefield.

Berkowitz, Eric. 2005. "The Subway Mayor." *LA Weekly*, August 19–25. Retrieved March 31, 2006 from http://www.laweekly.com.

Bijker, Wiebe, Thomas Hughes, and Trevor Pinch, eds. 1987. *The Social Construction of Technological Systems: New Directions in the Sociology and History of Technology*. MIT Press.

Bimber, Bruce, and David Guston. 1995. "Politics by the Same Means: Government and Science in the United States." In *Handbook of Science and Technology Studies*, ed. Sheila Jasanoff, Gerald E. Markle, James Peterson, and Trevor Pinch. Sage.

Blickstein, Susan, and Susan Hanson. 2001. "Critical Mass: Forging a Politics of Sustainable Mobility in the Information Age." *Transportation* 28 (4): 347–362.

Blumberg, Louis, and Robert Gottlieb. 1989. *War on Waste: Can America Win its Battle with Garbage?* Island.

Blumenstyk, Goldie. 2003. "Greening the World or 'Greenwashing' a Reputation?" *Chronicle of Higher Education*, January 10: A22–A24.

Bockmeyer, Janice. 2003. "Devolution and the Transformation of Community Housing Activism." *Social Science Journal* 40: 175–188.

Bohl, Charles. 2000. "New Urbanism and the City: Potential Applications and Implications for Distressed Inner-City Neighborhoods." *Housing Policy Debate* 11 (4): 761–801.

Böhme, Bernot, Wolfgang van den Daele, and Wolfgang Krohn. 1976. "Finalization in Science." *Social Science Information* 15: 307–330.

Bonanno, Alessandro, Lawrence Busch, William Friedland, Lourdes Gouveia, and Enzo Mingione, eds. 1994. *From Columbus to ConAgra: The Globalization of Food and Agriculture.* University Press of Kansas.

Bonser, Charles, Eugene McGregor, Jr., and Clinton Oster, Jr. 2000. *American Public Policy Problems.* Prentice-Hall.

Booth, Douglas. 1998. *The Environmental Consequences of Growth: Steady-State Economics as an Alternative to Ecological Decline.* Routledge.

Boström, Magnus. 2006. "Regulatory Credibility and Authority through Inclusiveness: Standardization Organizations in Cases of Eco-Labelling." *Organization* 13 (3): 345–367.

Boulder Independent Business Alliance. 2004. "What Is the Boulder Independent Business Alliance?" Retrieved March 31, 2006 from http://www.boulder-iba.org.

Bourdieu, Pierre. 1971. "Genèse et structure du champ religieux." *Revue Française de Sociologie* 12 (3): 295–334.

Bourdieu, Pierre. 1975. "The Specificity of the Scientific Field and the Social Conditions of the Progress of Reason." *Social Science Information* 14 (6): 19–47.

Bourdieu, Pierre. 1977. *Outline of a Theory of Practice.* Cambridge University Press.

Bourdieu, Pierre. 1982. *A Economia das Trocas Simbólicas.* Perspectiva.

Bourdieu, Pierre. 1988. *Homo Academicus.* Stanford University Press.

Bourdieu, Pierre. 1998. *Acts of Resistance: Against the Tyranny of the Market.* New Press.

Bourdieu, Pierre. 2001. *Science of Science and Reflexivity.* University of Chicago Press.

Boyd, Richard, Philip Gasper, and J. D. Trout. 1991. *The Philosophy of Science.* MIT Press.

Boyer, Roger. 1990. *The Regulation School: A Critical Introduction.* Columbia University Press.

Brady, Judy. 2003. "Corrupted Science: How Industry Corrupts Research." Breast Cancer Action, Newsletter 79 (November-December). Retrieved March 31, 2006 from http://www.bcaction.org.

Brand, Ralf. 2006. *Synchronizing Science and Technology with Human Behavior.* Earthscan.

Bratt, Rachel. 1997. "CDCs: Contributions Outweigh Contradictions. A Reply to Randy Stoecker." *Journal of Urban Affairs* 19 (1): 23–28.

Bressi, Todd. 2002. *The Seaside Debates: A Critique of the New Urbanism.* Rizzoli.

Brieger, Tracey. 2002. "Pesticide Action Network's First Twenty Years: An Interview with Monica Moore." *Global Pesticide Campaigner*, May: 18–21.

Broad, Robin, and Zahara Heckscher. 2003. "Before Seattle: The Historical Roots of the Current Movements against Corporate-Led Globalization." *Third World Quarterly* 24 (4): 713–728.

Bronfenbrenner, Kate, and Stephanie Luce. 2004. "The Changing Nature of Corporate Global Restructuring: The Impact of Production Shifts on Jobs in the U.S., China and Around the Globe." US-China Economic and Security Review Commission. Retrieved March 31, 2006 from http://digitalcommons.ilr .cornell.edu.

Brown, Allison. 2001. "Counting Farmers' Markets." *Geographical Review* 91 (4): 655–674.

Brown, Allison. 2002. "Farmers' Market Research 1940–2000: An Inventory and Review." *American Journal of Alternative Agriculture* 17 (4): 167–176.

Brown, Lester. 2001. *Eco-economy: Building an Economy for the Earth*. Norton.

Brown, M. Gordon. 2002. "*Charter of the New Urbanism, Congress of the New Urbanism*, Edited by Michael Leccese and Kathleen McCormick." *Journal of Real Estate Literature* 10 (1): 145–152.

Brown, Phil. 2007. *Contested Illnesses: Toward a New Environmental Health Movement*. Columbia University Press.

Brown, Phil, and Edwin Mikkelsen. 1990. *No Safe Place: Toxic Waste, Leukemia, and Community Action*. University of California Press.

Brown, Phil, and Stephen Zavestoski. 2004. "Social Movements in Health: An Introduction." *Sociology of Health and Illness* 26 (6): 679–694.

Brown, Phil, Stephen Zavestoski, and Sabrina McCormick. 2001. "A Gulf of Differences: Disputes over *Gulf War-Related* Illness." *Journal of Health and Social Behavior* 42 (3): 235–257.

Brown, Phil, Stephen Zavestoski, Sabrina McCormick, Brian Mayer, Rachel Morello-Frosch, and Rebecca Gasior. 2004. "Embodied Health Movements: Uncharted Territory in Social Movement Research." *Sociology of Health and Illness* 26: 1–31.

Brown, Phil, Stephen Zavestoski, Brian Mayer, Sabrina McCormick, and Pamela Webster. 2002. "Social Policy and Social Movements: Policy Issues in Environmental Health Disputes." *Annals of the American Academy of Political and Social Science* 584: 175–202.

Browne, William. 1983. "Mobilizing and Activating Group Demands: The American Agriculture Movement." *Social Science Quarterly* 64: 19–34.

Brulle, Robert, and Jonathan Essoka. 2005. "Whose Environmental Justice? An Analysis of the Governance Structure of Environmental Justice Organizations in the United States." In *Power, Justice and the Environment*, ed. David Pellow and Robert Brulle. MIT Press.

Brulle, Robert, and J. Craig Jenkins. 2006. "Spinning Our Way to Sustainability." *Organization and Environment* 19 (1): 82–87.

Brulle, Robert, and David Pellow. 2006. "Environmental Justice: Human Health and Environmental Inequalities." *Annual Review of Public Health* 27: 103–124.

Bryld, Erik. 2003. "Potentials, Problems, and Policy Implications for Urban Agriculture in Developing Countries." *Agriculture and Human Values* 20: 79–83.

Buck, Daniel, Christina Getz, and Julie Guthman. 1997. "From Farm to Table: The Organic Vegetable Commodity Chain of Northern California." *Sociologia Ruralis* 37 (1): 3–20.

Bullard, Robert. 1994. *Dumping in Dixie: Race, Class, and Environmental Quality*. Westview.

Bullard, Robert, Glenn Johnson, and Angel Torres. 2000a. "Dismantling Transportation Apartheid: The Quest for Equity." In *Sprawl City*, ed. Robert Bullard, Glenn Johnson, and Angel Torres. Island.

Bullard, Robert, Glenn Johnson, and Angel Torres. 2000b. "Environmental Costs and Consequences of Sprawl." In *Sprawl City*, ed. Robert Bullard, Glenn Johnson, and Angel Torres. Island.

Bullard, Robert, Glenn Johnson, and Angel Torres, eds. 2004. *Highway Robbery: Transportation Racism and New Routes to Equity*. South End.

Bullock, Simon. 2000. "The Economic Benefits of Farmers' Markets." Retrieved March 31, 2006 from http://www.foe.co.uk.

Busch, Lawrence. 1994. "The State of Agricultural Science and the Agricultural Science of the State." In *From Columbus to ConAgra*, ed. Alessandro Bonanno, Lawrence Busch, William Friedland, Lourdes Gouveia, and Enzo Mingione. University Press of Kansas.

Busch, Lawrence. 2000. "The Moral Economy of Grades and Standards." *Journal of Rural Studies* 16 (3): 273–283.

Busch, Lawrence, and William Lacy. 1981. "Sources of Influence on Problem Choice in the Agricultural Sciences: The New Atlantis Revisited." In *Science and Agricultural Development*, ed. Lawrence Busch. Allanheld, Osmun.

Business Alliance for Local Living Economies. 2006a. "About Us." Retrieved March 31, 2006 from http://www.livingeconomies.org.

Business Alliance for Local Living Economies. 2006b. "Marketplace." Retrieved March 31, 2006 from http://livingeconomies.org.

Bus Riders Union. 2004a. "A New Vision for Urban Transportation." Retrieved March 31, 2006 from http://www.busridersunion.org.

Bus Riders Union. 2004b. "Overview and History." Retrieved March 31, 2006 from http://www.busridersunion.org.

Buttel, Frederick. 2000a . "Classical Theory and Contemporary Environmental Sociology: Some Reflections on the Antecedents and Prospects for Reflexive Modernization Theories in the Study of Environment and Society." In *Environment and Global Modernity*, ed. Gert Spaargaren, Arthur Mol, and Frederick Buttel. Sage.

Buttel, Frederick. 2000b. "Ecological Modernization as a Social Theory." *Geoforum* 31 (1): 57–65.

Buttel, Frederick. 2006. "Ever Since Hightower: The Politics of Agricultural

Research in a Molecular Age." *Agriculture, Food, and Human Values* 22 (3): 275–283.

Buttel, Frederick, and Kenneth Gould. 2004. "Global Social Movement (s) at the Crossroads: Some Observations on the Trajectory of the Anti-Corporate Globalization Movement." *Journal of World Systems Research* 10 (1): 37–66.

Cahn, Edgar. 2004. "Keynote Address." Local Currencies in the Twentieth Century, conference at Bard College.

California Certified Organic Farmers. 2003. "About CCOF." Retrieved March 31, 2006 from http://www.ccof.org.

Callon, Michel. 1986. "Some Elements of a Sociology of Translation: Domestication of the Scallops and Fishermen." In *Power, Action, and Belief*, ed. John Law. Routledge.

Callon, Michel. 1994. "Four Models for the Dynamics of Science." In *Handbook of Science and Technology Studies*, ed. Sheila Jasanoff, Gerald Markle, James Peterson, and Trevor Pinch. Sage.

Calthorpe, Peter, and William Fulton. 2001. *The Regional City: Planning for the End of Sprawl.* Island.

Campbell, Donald. 1990. "Evolutionary Roles for Selection Theory." In *Evolution, Cognition, and Realism*, ed. Nicholas Rescher. University Press of America.

Campbell, Hugh, and Ruth Liepins. 2001. "Naming Organics: Understanding Organic Standards in New Zealand as a Discursive Field." *Sociologia Ruralis* 41 (1): 21–39.

Campbell, John. 1988. *Collapse of an Industry: Nuclear Power and the Contradictions of U.S. Policy*. Cornell University Press.

Campbell, Nancy. 2000. *Using Women: Gender, Drug Policy, and Social Justice*. Routledge.

Campbell, Scott. 1996. "Green Cities, Growing Cities, Just Cities? Urban Planning and the Contradictions of Sustainable Development." *American Planning Association Journal* 62 (3): 296–312.

Carman, Diane. 2006. "Plug Pulled on Renewable Energy Gurus." DenverPost.com, February 14. Retrieved March 31, 2006 from http://www.energybulletin.net.

Carr, James, and Zhong Yi Tong. 2002. "Introduction: Replicating Microfinance in the United States." In *Replicating Microfinance in the United States*, ed. James Carr and Zhong Yi Tong. Woodrow Wilson Center Press.

Carson, Lyn, and Brian Martin. 2002. "Random Selection of Citizens for Technological Decision Making." *Science and Public Policy* 29 (2): 105–113.

Carson, Rachel. 1962. *Silent Spring*. Houghton Mifflin.

Cassidy, Robert, ed. 2003a. "The Basics of LEED." *Building Design and Construction* November (supplement): 8–12.

Cassidy, Robert, ed. 2003b. "A Brief History of Green Building." *Building Design and Construction* November (supplement): 4–7.

Castells, Manuel. 1986. *The Rise of the Network Society*. Blackwell.

Castells, Manuel. 2001. *The Internet Galaxy: Reflections on Business, the Internet, and Society*. Oxford University Press.

Castells, Manuel, and Peter Hall. 1994. *Technopoles of the World: The Making of Twenty-First Century Industrial Complexes*. Routledge.

Chase-Dunn, Christopher, and Thomas Hall. 1997. *Rise and Demise: Comparing World Systems*. Westview.

Chefs Collaborative. 2005. "About Chefs Collaborative." Retrieved March 31, 2006 from http://www.chefscollaborative.org.

Chernow, Martin. 2002. "Introduction." *Bulletin Series: Yale School of Forestry and Environmental Studies* 106: 9–22.

Christen, Robert, and Deborah Drake. 2002. "Commercialization: The New Reality of Microfinance." In *The Commercialization of Microfinance*, ed. Deborah Drake and Elisabeth Rhyne. Kumarian.

Citizens Environmental Coalition and Sierra Club. 2005. *The Wasting of Rural New York State: Factory Farms and Public Health*. Citizens' Environmental Coalition.

Clarke, Adele. 1998. *Disciplining Reproduction: Modernity, American Life Sciences, and "the Problems of Sex."* University of California Press.

Clarke, Adele. 2000. "Maverick Reproductive Scientists and the Production of Contraceptives, 1915–2000+." In *Bodies of Technology*, ed. A. Saetnan, N. Oudshoorn, and M. Kirejczyk. Ohio State University Press.

Clarke, Adele, and Theresa Montini. 1993. "The Many Faces of RU486: Tales of Situated Knowledges and Technologial Contestations." *Science, Technology, and Human Values* 18 (1): 42–78.

Clarke, John. 2003. *Worlds Apart: Civil Society and the Battle for Ethical Globalization*. Kumarian.

Cockburn, Alexander, and Susanna Hecht. 1989. *Fate of the Forest: Developers, Destroyers, and Defenders of the Amazon*. Verso.

Cockburn, Cynthia. 1999. "The Material of Male Power." In *The Social Shaping of Technology*, ed. Donald MacKenzie and Judy Wajcman. Open University Press.

Cohen, Maurie. 2005. "Sustainable Consumption, American Style: Nutrition Education, Active Living, and Financial Literacy." *International Journal of Sustainable Development and World Ecology* 12 (4): 407–418.

Cohen, Maurie, and Joseph Murphy, eds. 2001. *Exploring Sustainable Consumption: Environmental Policy and the Social Sciences*. Pergamon.

Cohen, Robin, and Shirin Rai, eds. 2000. *Global Social Movements: Towards a Cosmopolitan Politics*. Athlone.

Cohousing Association of America. 2003. "Innovative Housing Concept Doubles in Size." Retrieved July 7, 2004 from http://www.cohousing.org.

Cole, Luke, and Sheila Foster. 2000. *From the Ground Up: Environmental Racism and the Rise of the Environmental Justice Movement*. New York University Press.

Collins, Harry. 1983. "An Empirical Relativist Program in the Sociology of Scientific Knowledge." In *Science Observed*, ed. Karin Knorr-Cetina and Michael Mulkay. Sage.

Collins, Harry. 1985. *Changing Order: Replication and Induction in Scientific Practice*. Sage.

Collins, Harry. 2000. "Surviving Closure: Post-rejection Adaptation and Plurality in Science." *American Sociological Review* 65 (6): 824–825.

Collins, Harry. 2002. "The Third Wave of Science Studies: Studies of Expertise and Experience." *Social Studies of Science* 32 (2): 235–296.

Collins, Randall, and Sal Restivo. 1983. "Robber Barons and Politicians in Mathematics: A Conflict Model for Science." *Canadian Journal of Sociology* 8 (2): 199–227.

Community Alliance with Family Farmers. 2003. "History." Retrieved March 31, 2006 from http://www.caff.org.

Community Involved in Sustaining Agriculture. 2005. "Community Involved in Sustaining Agriculture." Retrieved March 31, 2006 from http://www.buylocalfood.com.

Computer TakeBack. 2005. "About the Campaign." Retrieved March 31, 2006 from http://www.computertakeback.com.

Cone, Cynthia Abbott, and Andrea Myhre. 2000. "Community-Supported Agriculture: A Sustainable Alternative to Industrial Agriculture?" *Human Organization* 59 (2): 187–197.

Conford, Philip. 2000. *The Origins of the Organic Movement*. Floris Books.

Conford, Philip. 2002. "The Myth of Neglect: Responses to the Early Organic Movement, 1930–1950." *Agricultural History Review* 50 (1): 89–106.

Congress for the New Urbanism. 2006a. "Charter." Retrieved March 31, 2006 from http://www.cnu.org.

Congress for the New Urbanism. 2006b. "CNU History." Retrieved March 31, 2006 from http://www.cnu.org.

Connett, Paul, and Bill Sheehan. 2001. "Citizens' Agenda for Zero Waste: A U.S. /Canadian Perspective." Retrieved March 31, 2006 from http://www.grrn.org.

Consumer Reports. 2006. "Fighting for a Strong 'Organic' Label." *Consumer Reports*, February: 57.

Co-op America. 2003. "Co-op America Members Help Raise Wind Turbine." *Co-op America Quarterly* 60 (summer): 30.

Co-op America. 2006. "National Green Pages." Retrieved March 31, 2006 from http://www.greenpages.org.

Cooperman, David. 2003. "CESA Small Wind Project Overview Report." Clean Energy States Alliance. Retrieved January 20, 2005 from http://www.cleanenergystates.org.

Cordes, Helen. 2003. "Cotton-Picking Dynamo." *Utne Reader*, January-February: 48–51.

Cotugna, Nancy, and Patricia Dobbe Beebe. 2002. "Food Banking in the Twenty-First Century: Much More Than a Canned Handout." *Journal of the American Dietetic Association* 102: 1386–1388.

Cowen, Ruth Schwartz. 1983. *More Work for Mother: The Ironies of Household Technology from the Open Hearth to the Microwave.* Basic Books.

Cowen, Ruth Schwartz. 1999. "How the Refrigerator Got Its Hum." In *The Social Shaping of Technology,* ed. Donald MacKenzie and Judy Wajcman. Open University Press.

Crawford, Joel. 2000. *Carfree Cities.* International Books.

Credit Union National Association. 2003. "Annual Report." Retrieved March 31, 2006 from http://www.cuna.org.

Creech, Dennis, and Natalie Brown. 2000. "Energy Use and the Environment." In *Sprawl City,* ed. Robert Bullard, Glenn Johnson, and Angel Torres. Island.

Croissant, Jennifer, and Sal Restivo, eds. 2001. *Degrees of Compromise: Industrial Interests and Academic Values.* SUNY Press.

Curtis, Karen. 1997. "Urban Poverty and the Social Consequences of Privatized Food Assistance." *Journal of Urban Affairs* 19 (2): 207–226.

Daley-Harris, Sam. 2002. *Pathways Out of Poverty: Innovations in Microfinance for the Poorest Families.* Kumarian.

Danaher, Kevin, and Jason Mark. 2003. *Insurrection: Citizen Challenges to Corporate Power.* Routledge.

Daviron, Benoit, and Stefano Ponte. 2005. *The Coffee Paradox: Global Markets, Commodity Trade and the Elusive Promise of Development.* Zed Books.

Della Porta, Donatella, and Sidney Tarrow, eds. 2005. *Transnational Protest and Global Activism: People, Passions, and Power.* Rowman and Littlefield.

Demerath, Loren, and David Levinger. 2003 "The Social Qualities of Being on Foot: A Theoretical Analysis of Pedestrian Activity, Community, and Culture." *City and Community* 2 (3): 217–237.

DeSimone, Livio, Frank Popoff, with the World Business Council for Sustainable Development. 1997. *Eco-Efficiency: The Business Link to Sustainable Development.* MIT Press.

DeSoucey, Michaela. 2004. "Slow, Close, and Pure: Constructing Virtuous Food as Integrated Social Movements and Marketplace." Paper presented at annual meeting of Agriculture, Food, and Human Values Society, Hyde Park, N.Y.

Desrochers, Pierre. 2001. "Eco-Industrial Parks: The Case for Private Planning." *Independent Review* 5 (3): 345–372.

Diamond, Jarred. 2005. *Collapse: How Societies Choose to Fail or Succeed.* Viking.

Dobb, Edwin. 2000. "Growing Resistance: Farm-Raised Killer Microbes." *Mother Jones* 25 (6): 3.

Dowie, Mark. 1995. *Losing Ground: American Environmentalism at the End of the Century.* MIT Press.

Dow Jones Sustainability Indexes. 2005. "Welcome to the Dow Jones Sustainability Indexes." Retrieved March 31, 2006 from http://www.sustainability-indexes.com.

Downey, Gary. 1988. "Structure and Practice in the Cultural Identities of Scientists: Negotiating Nuclear Wastes in New Mexico." *Anthropological Quarterly* 61 (1): 26–38.

Dreier, Peter. 1997. "The New Politics of Housing: How to Build a Constituency for a Progressive Federal Housing Policy." *Journal of the American Planning Association* 63 (winter): 5–27.

Drucker, Peter. 1993. *Post-Capitalist Society*. HarperCollins.

Duhem, Pierre. 1982. *The Aim and Structure of Physical Theory*. Princeton University Press.

Dumont, Louis. 1977. *From Mandeville to Marx: The Genesis and Triumph of Economic Ideology*. University of Chicago Press.

Dumont, Louis. 1986. *Essays on Individualism: Modern Ideology in Anthropological Perspective*. University of Chicago Press.

Dupuis, E. Melanie. 2000. "Not in My Body: BGH and the Rise of Organic Milk." *Agriculture and Human Values* 17 (3): 285–295.

Dupuis, E. Melanie. 2002. *Nature's Perfect Food: How Milk Became America's Drink*. New York University Press.

Dupuis, E. Melanie, ed. 2004. *Smoke and Mirrors: The Politics and Culture of Air Pollution*. New York University Press.

Durkheim, Emile. 1964. *The Division of Labor in Society*. Macmillan/Free Press.

Earth Island Institute. 2004. "International Marine Mammal Project: Bush Administration Wrong on Dolphin Protection." *Earth Island Journal* 19 (3): n.p. Retrieved March 31, 2006 from http://www.earthisland.org.

EcoCity Cleveland. 2004. "High-Performance Town Homes for the Cleveland EcoVillage." Retrieved March 31, 2006 from http://www.ecocitycleveland.org.

Economics Institute. 1999. "Catalysts for Growth: Farmers Markets as a Stimulus for Economic Development." Retrieved March 31, 2006 from http://www.loyno.edu.

Ecovillage Network of the Americas. 2006. "ENA: Ecovillage Network of the Americas." Retrieved March 31, 2006 from http://ena.ecovillage.org.

Edelman, Mark. 2005. "Bringing the Moral Economy Back in . . . to the Study of 21st-Century Transnational Peasant Movements." *American Anthropologist* 107 (3): 331–345.

Efrati, Amir. 2006. "EBay Sellers Reshape Garage-Sale Dynamic." The *Wall Street Journal* Center for Entrepreneurs. Retrieved April 27, 2006 from http://www.startupjournal.com.

Eglash, Ron, Jen Croissant, Giovanna Di Chiro, and Ray Fouche, eds. 2004. *Appropriating Technology: Vernacular Science and Social Power*. University of Minnesota Press.

Eisinger, Peter. 1998. *Toward an End to Hunger in America*. Brookings Institution. Retrieved March 31, 2006 from http://brookings.nap.edu.

Elliott, James, Kevin Gotham, and Melinda Milligan. 2004. "Framing the Urban: Struggles over HOPE VI and New Urbanism in a Historic City." *City and Community* 3 (4): 373–394.

Engell, James, and Anthony Dangerfeld. 1998. "The Market-Model University." *Harvard Magazine*, May-June: 48–55, 111.

EnviroWindows. 2002. "Organic Farming in Europe: Recent Developments and Future Prospects." Retrieved March 31, 2006 from http://ewindows.eu.org.

Epstein, Steven. 1996. *Impure Science: AIDS, Activism, and the Politics of Knowledge*. University of California Press.

Etc. Group. 2004a. "Nano's Troubled Waters." Retrieved February 21, 2005 from http://www.etcgroup.org.

Etc. Group. 2004b. "Nanotech: Unpredictable and Un-regulated." Retrieved February 21, 2005 from http://www.etcgroup.org.

Etzkowitz, Henry, and Loet Leydesdorff, eds. 1997. *Universities in the Global Economy: A Triple Helix of University-Industry-Government Relations*. Pinter.

Etzkowitz, Henry, and Loet Leydesdorff. 1999. "The Future Location of Research and Technology Transfer." *Journal of Technology Transfer* 24 (2-3): 111–123.

Etzkowitz, Henry, Andrew Webster, and Peter Healy. 1998. *Capitalizing Knowledge: New Intersections of Industry and Academia*. SUNY Press.

Evans, Peter, ed. 2002. *Livable Cities? Urban Struggles for Livelihood and Sustainability*. University of California Press.

Farkus, Nicole. 2002. Bread, Cheese, and Expertise: Dutch Science Shops and Democratic Institutions. Ph.D. dissertation, Rensselaer Polytechnic Institute.

Feenberg, Andrew. 1995. *Alternative Modernity: The Technical Turn in Philosophy and Social Theory*. University of California Press.

Feenstra, Gail, Christopher Lewis, C. Clare Hinrichs, Gilbert Gillespie, and Duncan Hilchey. 2003. "Entrepreneurial Outcomes and Enterprise Size in U.S. Retail Farmers' Markets." *American Journal of Alternative Agriculture* 18 (1): 46–55.

Fenn, Paul. 2004. "San Francisco Declares 'Energy Independence.'" Retrieved March 31, 2006 from http://www.local.org.

Firebaugh, Glenn. 2003. *The New Geography of Global Income Inequality*. Harvard University Press.

Fischer, Frank. 1995. *Evaluating Public Policy*. Nelson-Hall.

Fischer, Frank. 2000. *Citizens, Experts, and the Environment: The Politics of Local Knowledge*. Duke University Press.

Fitzgerald, Jenrose. 2005. Citizens, Experts, and the Economy: Struggles over Globalization and the "New Economy" in Kentucky. Ph.D. dissertation, Rensselaer Polytechnic Institute.

Flaccavento, Anthony. 2002. "From the Earth, Up." *Yes! A Journal of Positive Futures*, fall: 25–27.

Flacks, Richard. 2004. "Knowledge for What? Thoughts on the State of Social Movement Studies." In *Rethinking Social Movements*, ed. Jeff Goodwin and James Jasper. Rowman and Littlefield.

Florida, Richard. 1996. "Lean and Green: The Move to Environmentally Conscious Manufacturing." *California Management Review* 39 (1): 80–105.

Forman, Paul. 1987. "Behind Quantum Electronics: National Security as a Basis for Physical Research in the U.S., 1940–1960." *Historical Studies in the Physical Sciences* 18: 149–229.

Forsythe, Diana. 2001. *Studying Those Who Study Us: An Anthropologist in the World of Artificial Intelligence*. Stanford University Press.

Foucault, Michel. 1980. *Power/Knowledge: Selected Interviews and Other Writings, 1972–1977*, ed. Colin Gordon. Pantheon.

Franklin, Sarah, and Margaret Lock, eds. 2003. *Remaking Life and Death: Toward an Anthropology of the Biosciences*. School of American Research Press.

Freire, Paulo. 1986. *Pedagogy of the Oppressed*. Continuum.

Freudenburg, William. 2005. "Privileged Access, Privileged Accounts: Toward a Socially Structured Theory of Resources and Discourses." *Social Forces* 84 (1): 89–114.

Frickel, Scott. 2004a. *Chemical Consequence: Environmental Mutagens, Scientist Activism, and the Rise of Genetic Toxicology*. Rutgers University Press.

Frickel, Scott. 2004b. "Just Science? Organizing Scientist Activism in the U.S. Environmental Justice Movement." *Science as Culture* 13 (4): 449–471.

Frickel, Scott, and Neil Gross. 2005. "A General Theory of Scientific/Intellectual Movements." *American Sociological Review* 70: 204–232.

Frickel, Scott, and Kelly Moore. 2006. *The New Political Sociology of Science: Institutions, Networks, and Power*. University of Wisconsin Press.

Friedman, Jonathan. 2003. "Globalization, Disintegration, Reorganization: The Transformation of Violence." In *Globalization, the State, and Violence*, ed. Jonathan Friedman. Altamira.

Friedman, Monroe. 1999. *Consumer Boycotts: Effecting Change through the Marketplace and the Media*. Routledge.

Friedman, Thomas. 1999. *The Lexus and the Olive Tree: Understanding Globalization*. Farrar, Straus, and Giroux.

Friedman, Thomas. 2005. *The World Is Flat: A Brief History of the Twenty-First Century*. Farrar, Straus, and Giroux.

Friends of the Earth. 2006. "Oil, Mining, and Gas: Climate Change Litigation." Retrieved March 31, 2006 from http://www.foe.org.

Fujimura, Joan. 1987. "Constructing Doable Problems in Cancer Research: Articulating Alignment." *Social Studies of Science* 17: 257–293.

Fuller, Steve. 2000a. *The Governance of Science: Ideology and the Future of the Open Society*. Open University Press.

Fuller, Steve. 2000b. *Thomas Kuhn: A Philosophical History for Our Times*. University of Chicago Press.

Fulton, William. 1996. "The New Urbanism Challenges Conventional Planning." *Land Lines Newsletter* 8 (5), September, n.p. Retrieved March 31, 2006 from http://www.lincolninst.edu.

Furniture Bank. 2006. "Furniture Bank: Our Mission." Retrieved March 31, 2006 from http://www.furniturebank.org.

Galayda, Jaime Radesi. 2006. Community-Supported Agriculture: Social Preferences and Well-Being. Ph.D. dissertation, Rensselaer Polytechnic Institute.

Gamson, William. 1990. *The Strategy of Social Protest*. Wadsworth.

Gans, Herbert. 1959. "The Human Implications of Current Redevelopment and Relocation Planning." *Journal of the American Institute of Planners* 25 (1): 15–25.

Gerefi, Gary, Ronie Garcia-Johnson, and Erika Sasser. 2001. "The NGO-Industrial Complex." *Foreign Policy* 125 (July-August): 56–65.

Gibbons, Michael, Camille Limoges, Helga Nowotny, Simon Schwartzman, Peter Scott, and Martin Trow. 1994. *The New Production of Knowledge: The Dynamics of Science and Research in Contemporary Societies*. Sage.

Gibbs, Lois. 2002. "Social Policy and Social Movements: Citizen Activism for Environmental Health. The Growth of a Powerful New Grassroots Health Movement." *Annals of the American Academy of Political and Social Science* 584: 97–109.

Giddens, Anthony. 1995. *Politics, Sociology, and Social Theory: Encounters with Classical and Contemporary Social Thought*. Stanford University Press.

Gillan, Stuart, and Laura Starks. 2000. "Corporate Governance Proposals and Shareholder Activism: The Role of Institutional Investors." *Journal of Financial Economics* 57 (2): 275–305.

Gillan, Stuart, and Laura Starks. 2003. "Corporate Governance, Corporate Ownership, and the Role of Institutional Investors: A Global Perspective." *Journal of Applied Finance* 13 (2): 4–22.

Gillham, Oliver. 2002. *The Limitless City: A Primer on the Urban Sprawl Debate*. Island.

Gipe, Paul. 1995. *Wind Energy Comes of Age*. Wiley.

Gipe, Paul. 2002. *Small Wind Turbines Sprouting as Power Prices Rise*. Retrieved March 31, 2006 from http://www.wind-works.org.

Gish, Melinda. 2001. "94-211: The Low Income Home Energy Assistance Program." CRS Report for Congress. Retrieved March 31, 2006 from http://www.ncseonline.org.

Giugni, Marco. 1998. "Was It Worth the Effort? The Outcomes and Consequences of Social Movements." *Annual Review of Sociology* 24: 371–393.

Giugni, Marco, Doug McAdam, and Charles Tilly, eds. 1999. *How Social Movements Matter*. University of Minnesota Press.

Global Ecovillage Network. 2005a. "GEN Mission, Vision, and Purposes." Retrieved March 31, 2006 from http://gen.ecovillages.org.

Global Ecovillage Network. 2005b. "The History of GEN." Retrieved March 31, 2006 from http://gen.ecovillages.org.

Global Green USA. 2006. "Green Affordable Housing Initiative." March 31, 2006 from http://www.globalgreen.org.

GMWatch. 2006. "Articles Archive." Retrieved March 31, 2006 from http://www.gmwatch.org.

Gofman, John, and Egan O'Connor. 1985. *X-Rays: Health Effects of Common Exams*. Sierra Club Books.

Goldberg, Jessica. 2001. "Research Orientation and Sources of Influence: Agricultural Scientists in the U.S. Land-Grant System." *Rural Sociology* 66 (1): 69–92.

Goodman, David. 2000. "The Changing Bio-politics of the Organic: Production, Regulation, Consumption." *Agriculture and Human Values* 17 (3): 211–213.

Goodman, David, and Melanie Dupuis. 2000. "Knowing Food and Growing Food: Beyond the Production-Consumption Debate in the Sociology of Agriculture." *Sociolgia Ruralis* 42 (1): 6–23.

Goodman, David, and Michael Goodman. 2001. "Sustaining Foods: Organic Consumption and the Socio-ecological Imaginary." In *Exploring Sustainable Consumption*, ed. Maurie Cohen and Joseph Murphy. Elsevier.

Goodman, Richard. 2000. *Report on Community Gardening*. National Gardening Association.

Gorman, Michael, and Matthew Mehalik. 2002. "Turning Good into Gold: A Comparative Study of Two Environmental Networks." *Science, Technology, and Human Values* 27 (4): 499–529.

Gottlieb, Robert. 1993. *Forcing the Spring: The Transformation of the American Environmental Movement*. Island.

Gottlieb, Robert. 2001. *Environmentalism Unbound: Exploring New Pathways for Change*. MIT Press.

Gottlieb, Robert, and Andrew Fisher. 1996. "Community Food Security and Environmental Justice: Searching for a Common Discourse." *Agriculture and Human Values* 3 (3): 23–32.

Gould, Kenneth Allan Schnaiberg, and Adam Weinberg. 1996. *Local Environmental Struggles: Citizen Activism and the Treadmill of Production*. Cambridge University Press.

Gould, Kenneth, Tammy Lewis, and J. Timmons Roberts. 2004. "Blue-green Coalitions: Constraints and Possibilities in the Post 9-11 Political Environment." *Journal of World Systems Research* 10 (1): 91–116.

Government Purchasing Project. 2005. "Government Programs on EPP." Retrieved March 31, 2006 from http://www.gpp.org.

Grace Factory Farm Project. 2006. "Resources: Regulations and Legislation: Moratoria on Factory Farms." Retrieved March 31, 2006 from http://www .factoryfarm.org.

Grassroots Recycling Network. 2001. "Annual Report." Retrieved March 31, 2006 from http://www.grrn.org.

Grassroots Recycling Network. 2002. "Annual Report." Retrieved March 31, 2006 from http://www.grrn.org.

Greco, Thomas. 2001. *Money: Understanding and Creating Alternatives to Legal Tender*. Chelsea Green.

Green Affordable Housing Coalition. 2004. "Green Affordable Housing Coalition." Retrieved March 31, 2006 from http://frontierassoc.net.

Green Institute. 2003. "The Green Institute." Retrieved March 31, 2006 from http://www.greeninstitute.org.

Green, Lawrence, and Shawna Mercer. 2001. "Can Public Health Researchers and Agencies Reconcile the Push from Funding Bodies and the Pull from Communities?" *American Journal of Public Health* 91 (12): 1926–1929.

Greene, Wade. 1971. "J. I. Rodale: Guru of the Organic Food Cult." *New York Times Magazine*, June 6: 30–31, 54–60, 65, 68.

Greenpeace. 2004. "Victory! Monsanto Drops Roundup Ready Wheat." Retrieved March 31, 2006 from http://www.greenpeace.org.

Green Restaurant Association. 2002a. "About Us." Retrieved March 31, 2006 from http://www.dinegreen.com.

Green Restaurant Association. 2002b. "Environmental Guidelines." Retrieved March 31, 2006 from http://www.dinegreen.com.

Greenwood, Davydd, and Morten Levin. 1998. *Introduction to Action Research: Social Research for Social Change*. Sage.

Gregson, Nicky, and Louise Crewe. 2003. *Second-Hand Cultures*. Berg.

Grimes, Kimberly. 2000. "Democratizing International Production and Trade: North American Alternative Trading Organizations." In *Artisans and Cooperatives*, ed. Kimberly Grimes and Lynne Milgram. University of Arizona Press.

Grimes, Kimberly. 2005. "Changing the Rules of Trade with Global Partnerships: The Fair Trade Movement." In *Social Movements*, ed. June Nash. Blackwell.

Griscom, Amanda. 2001. "Power to the People: Company Business and Marketing." Retrieved March 31, 2006 from http://www.findarticles.com.

Guardian. 2000. "Fertile Minds." April 26, n.p. Retrieved March 31, 2006 from http://www.guardian.co.uk.

Gunn, Christopher. 2004. *Third-Sector Development: Making Up for the Market*. Cornell University Press.

Gunther, Marc. 2006. "The Green Machine." *Fortune*, August 7: 42–57.

Gusterson, Hugh. 1996. *Nuclear Rites: A Weapons Laboratory at the End of the Cold War*. University of California Press.

Guston, David. 2001. "Boundary Organizations in Environmental Policy and Science: An Introduction." *Science, Technology, and Human Values* 26 (4): 399–408.

Guterman, Lila. 2000. "'Green Chemistry' Movement Seeks to Reduce Hazardous Byproducts of Chemical Processes." *Chronicle of Higher Education,* August 4: A17–A18.

Guthman, Julie. 1998. "Regulating Meaning, Appropriating Nature: The Codification of California Organic Agriculture." *Antipode* 30 (2): 135–154.

Guthman, Julie. 2000. "Raising Organic: An Agro-ecological Assessment of Grower Practices in California." *Agriculture and Human Values* 17 (3): 257–266.

Guthman, Julie. 2002. "Commodified Meanings, Meaningful Commodities: Rethinking Production-Consumption Links through the Organic System of Provision." *Sociologia Ruralis* 42 (4): 295–311.

Guthman, Julie. 2004. *Agrarian Dreams: The Paradox of Organic Farming in California.* University of California Press.

Haberern, John. 1997. "J. I. Rodale and the Rodale Family." Retrieved March 31, 2006 from http://www.depweb.state.pa.us.

Habermas, Jürgen. 1987. *A Theory of Communicative Action.* Volume 2. Beacon

Habitat for Humanity International. 2006a. "Environmental Initiative." Retrieved March 31, 2006 from http://www.habitat.org.

Habitat for Humanity International. 2006b. "Habitat for Humanity Fact Sheet." Retrieved March 31, 2006 from http://www.habitat.org.

Hadwinger, Don. 1993. "Henry Wallace: Champion of a Durable Agriculture." *American Journal of Alternative Agriculture* 8 (1): 2–3.

Hagstrom, Warren. 1964. "Anomie in Scientific Communities." *Social Problems* 12 (2): 186–195.

Hagstrom, Warren. 1965. *The Scientific Community.* Basic Books.

Halfman, Jost. 1999. "Community and Life Chances: Risk Movements in the United States and Germany." *Environmental Values* 8: 177–197.

Hall, Peter. 1992. *Inventing the Nonprofit Sector and Other Essays on Philanthropy, Voluntarism, and Nonprofit Organizations.* Johns Hopkins University Press.

Hamlett, Patrick. 2002. "Trust, Accountability, and Citizen Participation: Implementing the Danish Consensus Conference on the Internet." Paper presented at annual meeting of European Association of Studies of Science and Technology, University of York. Retrieved March 31, 2006 from http://www.york.ac.uk.

Haraway, Donna. 1989. *Primate Visions: Gender, Race, and Nature in the World of Modern Science.* Routledge.

Haraway, Donna. 1997. *Modest_Witness@Second_Millennium.FemaleMan© _Meets_OncoMouseTM.* Routledge.

Hård, Mikael, and Andrew Jamison. 2005. *Hubris and Hybrids: A Cultural History of Technology and Science.* Routledge.

Harding, Sandra. 1992. "After the Neutrality Idea: Science, Politics, and 'Strong Objectivity.'" *Social Research* 59 (3): 567–587.

Harding, Sandra, ed. 1993. *The "Racial" Economy of Science: Toward a Democratic Future*. Indiana University Press.

Harding, Sandra. 1998. *Is Science Multicultural? Postcolonialisms, Feminisms, and Epistemologies*. Indiana University Press.

Hargens, Lowell, and Diane Felmlee. 1984. "Structural Determinants of Stratification in Science." *American Sociological Review* 49: 685–697.

Harper, Krista. 2004. "The Genius of a Nation versus the Gene-Tech of a Nation: Science, Identity, and Genetically Modified Food in Hungary." *Science as Culture* 13 (4): 471–492.

Harvey, David. 1989. *The Condition of Postmodernity: An Enquiry into the Origins of Cultural Change*. Blackwell.

Harvey, David. 2000. *Spaces of Hope*. University of California Press.

Hassanein, Neva. 1999. *Changing the Way American Farms: Knowledge and Community in the Sustainable Agriculture Movement*. University of Nebraska Press.

Hawken, Paul, Amory Lovins, and L. Hunter Lovins. 1999. *Natural Capitalism: Creating the Next Industrial Revolution*. Little, Brown, and Co.

Hayes, Denis. 1979. *Blueprint for a Solar America*. Solar Lobby.

Heiman, Michael. 1990. "From 'Not In My Backyard!' to 'Not In Anybody's Backyard!': Grassroots Challenge to Hazardous Waste Facility Siting." *Journal of the American Planning Association* 56 (3): 359–362.

Heimlich, Ralph and William Anderson. 2001. "Development at and Beyond the Urban Fringe: Impacts on Agriculture." *Agricultural Outlook*, August: 15–18.

Heims, Steve. 1991. *The Cybernetics Group: Constructing a Social Science for Postwar America*. MIT Press.

Held, David. 1996. *Democracy and the Global Order: From the Modern State to Cosmopolitan Governance*. Stanford University Press.

Henderson, Hazel. 1996. *Creating Alternative Futures: The End of Economics*. Kumarian.

Henderson, Kathryn. 2003. "Straw-Bale Building: Using an Old Technology to Preserve the Environment." In *Inventing for the Environment*, ed. Arthur Molella and Joyce Bedi. MIT Press.

Henderson, Kathryn. 2006. "Ethics, Culture, and Structure in the Negotiation of Straw-Bale Building Codes." *Science, Technology, and Human Values*, forthcoming.

Herrmann, Gretchen. 1997. "Gift or Commodity: What Changes Hands in the U.S. Garage Sale?" *American Ethnologist* 24 (4): 910–930.

Hess, David. 1995. *Science and Technology in a Multicultural World: The Cultural Politics of Facts and Artifacts*. Columbia University Press.

Hess, David. 1997. *Science Studies: An Advanced Introduction*. New York University Press.

Hess, David. 1998. "The Problem of Undone Science." Paper presented at annual meeting of Society for Social Studies of Science, Halifax.

Hess, David. 1999. *Evaluating Alternative Cancer Therapies: A Guide to the Science and Politics of an Emerging Medical Field* . Rutgers University Press.

Hess, David. 2001. *Selecting Technology, Science, and Medicine. Alternative Pathways in Globalization*, volume 1. Retrieved March 31, 2006 from http://www.davidjhess.org.

Hess, David. 2002a. "The Raw and the Organic: Politics of Therapeutic Cancer Diets in the U.S." *Annals of the Academy of Political and Social Science* 583 (September): 76–97.

Hess, David. 2002b. "Stronger versus Weaker Integration Policies." *American Journal of Public Health* 92 (10): 1579–1581.

Hess, David. 2004a. "CAM Cancer Therapies in Twentieth-Century North America: Examining Continuities and Change." In *The Politics of Healing*, ed. Robert Johnston. Routledge.

Hess, David. 2004b. "The Green Technopole and Green Localism: Comparing Regional Development Strategies." Retrieved March 31, 2006 from http://www.davidjhess.org.

Hess, David. 2004c. "Organic Food and Agriculture in the U.S. : Object Conflicts in a Health-Environmental Movement." *Science as Culture* 13 (4): 493–513.

Hess, David. 2005a. "Medical Modernisation, Scientific Research Fields, and the Epistemic Politics of Health Social Movements." In *Social Movements in Health*, ed. Phil Brown and Stephen Zavestoski. Blackwell.

Hess, David. 2005b. "Technology and Product Oriented Movements: Approximating Social Movement Studies and Science and Technology Studies." *Science, Technology, and Human Values* 30 (4): 515–535.

Hess, David. 2006a. "Angiogenesis and the Dynamics of Scientific Fields." In *The New Political Sociology of Science*, ed. Scott Frickel and Kelly Moore. University of Wisconsin Press.

Hess, David. 2006b. "What Is a Clean Bus? Object Conflicts and the Greening of Public Transit." Paper presented at annual meeting of American Sociological Association.

Hess, David, Steve Breyman, Nancy Campbell, and Brian Martin. 2007. "Science, Technology, and Social Movements." Forthcoming in *Handbook of Science and Technology Studies*, ed. Edward Hackett, Olga Amsterdamska, Michael Lynch, and Judy Wajcman. MIT Press.

Hess, David, and Langdon Winner. 2005. "Case Studies of Sustainability, the Politics of Design, and Localism." Retrieved March 31, 2006 from http://www.davidjhess.org.

Hessen, Boris. 1971. *The Social and Economic Roots of Newton's Principia*. Howard Fertig.

Hickman, Lanier. 1999. "Garbage: Bin There, Done That." *American City and County* 114 (13): 60ff.

Hilgartner, Stephen. 2001. "Election 2000 and the Production of the Unknowable." *Social Studies of Science* 31 (3): 439–441.

Hirst, Paul. 2000. "The Global Economy: Myths or Reality?" In *The Ends of Globalization*, ed. Don Kalb, Marco van der Land, Richard Staring, Bart van Steenbergen, and Nico Wilterdink. Rowman and Littlefield.

Hirst, Paul, and Graham Thompson. 1999. *Globalization in Question: The International Economy and the Possibilities of Governance*, second edition. Blackwell/Oxford.

Hoffman, Lily. 1989. *The Politics of Knowledge: Activist Movements in Medicine and Planning*. SUNY Press.

Holdren, John. 1998. "Federal Energy Research and Development for the Challenges of the 21st Century." In *Investing in Innovation*, ed. Lewis Branscomb and James Keller. MIT Press.

Hollender, Jeffrey, and Stephen Fenichell. 2004. *What Matters Most: How a Small Group of Pioneers Is Teaching Social Responsibility to Big Business, and Why Big Business Is Listening*. Basic Books.

Home Power. 2006a. "Circulation and Audience." Retrieved March 31, 2006 from http://www.homepower.com.

Home Power. 2006b. "Guerilla solar rogues gallery." Retrieved January 9, 2004 from http://www.homepower.com.

Horne, Suzanne, and Avril Maddrell. 2002. *Charity Shops: Retailing, Consumption, and Society*. Routledge.

Hott, Lawrence. 1997. *Divided Highways: The Interstates and the Transformation of American Life*. Public Broadcasting System.

Howard, Ebenezer, and Frederic Osborn. 1898. *Garden Cities of Tomorrow*. Faber and Faber Ltd.

Howard, Jeffrey. 2004. Toward Intelligent, Democratic Steering of Chemical Technologies. Ph.D. dissertation, Rensselaer Polytechnic Institute.

Howard, Phil. 2006. "Who Owns What: Organic Industry Structure." Retrieved March 31, 2006 from http://www.certifiedorganic.bc.ca.

Huet, Tim. 2001. "News from Mondragon." *GEO* 30: 1–7. Retrieved March 31, 2006 from http://www.geonewsletter.org.

Hughes, Thomas. 1983. *Networks of Power: Electrification in Western Society, 1830–1900*. Johns Hopkins University Press.

Hughes, Thomas. 1987. "The Evolution of Large Technological Systems." In *The Social Construction of Technological Systems*, ed. Wiebe Bijker, Thomas Hughes, and Trevor Pinch. MIT Press.

Hung, Chi-Kan Richard. 2002. "From South to North: A Comparative Study of Group-Based Microcredit Programs in Developing Countries and the United States." In *Replicating Microfinance in the United States*, ed. James Carr and Zhong Yi Tong. Woodrow Wilson Center Press.

Inda, Jonathan, and Renato Rosaldo, eds. 2001. *The Anthropology of Globalization: A Reader*. Blackwell/Oxford.

Institute for Local Self-Reliance. 2006. "California Wants Biofuels Production to be Within State." New Rules Project, May 10. Retrieved May 12, 2006 from http://www.newrules.org.

International Council for Local Environmental Initiatives (ICLEI). n.d. "Case Study #3: Solid Waste Management. Bandung, Indonesia." ICLEI.

International Council for Local Environmental Initiatives (ICLEI). n.d. "Case Study #21: Community-Based Service Delivery. Quito, Ecuador." ICLEI.

International Council for Local Environmental Initiatives (ICLEI). n.d. "Case Study #60: Waste Management through Community Participation. Municipal Administration of Belo Horizonte, Brazil." ICLEI.

Irwin, Alan, and Brian Wynne, eds. 1996. *Misunderstanding Science? The Public Reconstruction of Science and Technology*. Cambridge University Press.

Israel, Barbara, Amy Schulz, Edith Parker, and Adam Becker. 1998. "Review of Community-based Research: Assessing Partnership Approaches to Improve Public Health." *Annual Review of Public Health* 19: 173–202.

Issel, William. 1999. "Land Values, Human Values, and the Preservation of the City's Treasured Appearance." *Pacific Historical Review* 68 (4): 611–646.

Jackson, Carlton. 1974. *J. I. Rodale: Apostle of Non-Conformity*. Pyramid Books.

Jackson, Hildur, and Ross Jackson. 2004. "Global Eco-village History, 1990–2004." Gaia Trust. Retrieved March 31, 2006 from http://www.gaia.org.

Jackson, Hildur, and Karen Svensson. n.d. *Ecovillage Living*. Green Books.

Jackson, Ross. 2004. "The Ecovillage Movement." *Permaculture Magazine* 40 (summer): n.p. Retrieved March 31, 2006 from http://www.gaia.org.

Jacob, Diane. 2003. "The Fine Art of Feeding the Hungry." *Gastronomica* 3 (4): 14–20.

Jacob, Jeffrey. 1997. *New Pioneers: The Back-to-the-Land Movement and the Search for a Sustainable Future*. Pennsylvania State University Press.

Jacob, Margaret. 1988. *The Cultural Meaning of the Scientific Revolution*. Knopf.

Jacobs, Jane. 1961. *The Death and Life of Great American Cities*. Vintage Books.

Jacobs, Jane. 1969. *The Economy of Cities*. Vintage.

Jain, Sarah. 2006. *Injury: The Politics of Product Design and Safety Law in the United States*. Princeton University Press.

Jamison, Andrew. 2001. *The Making of Green Knowledge: Environmental Politics and Cultural Transformation*. Cambridge University Press.

Jamison, Andrew, Ron Eyerman, Jacqueline Cramer, and Jeppe Laessøe. 1990. *The Making of the New Environmental Consciousness: A Comparative Study of Environmental Movements in Sweden, Denmark, and the Netherlands*. Edinburgh University Press.

Johansson, Phillip. 1999. "Calvert Brings Vanguard into Socially Responsible Arena." SocialFunds.com. Retrieved March 31, 2006 from http://www .socialfunds.com.

Joppke, Christian. 1993. *Mobilizing against Nuclear Energy: A Comparison of Germany and the United States*. University of California Press.

Jørgensen, Ulrik, and Peter Karnøe. 1996. "The Danish Wind-Turbine Story: Technical Solutions to Political Visions?" In *Managing Technology in Society*, ed. Arie Rip, Tom Misa, and Johan Schot. Pinter.

Jørgensen, Ulrik, and Lars Strunge. 2002. "Restructuring the Power Arena in Denmark: Shaping Markets, Technology, and Environmental Priorities." In *Shaping Technology, Guiding Policy*, ed. Knut Sørensen and Robin Williams. Elgar.

Journal of Political Ecology. 2006. "Journal of Political Ecology: Case Studies in Society and History." Retrieved March 31, 2006 from http://dizzy.library .arizona.edu.

Jurik, Nancy. 2005. *Bootstrap Dreams: U.S. Microenterprise Development in an Era of Welfare Reform*. ILR Press/Cornell University Press.

Kadet, Anne. 2005. "The Instant Small Town." *Smart Money*, August: 71–75.

Kaiser, Mark, and Allan Pulsipher. 2002. "LIHEAP Reconsidered." *Energy Policy* 31 (14): 1441–1458.

Kaltoft, Pernille. 2001. "Organic Farming in Late Modernity: At the Frontier of Modernity or Opposing Modernity?" *Sociologia Ruralis* 41 (1): 146–158.

Karr, Rick. 2005. "Prometheus Unbound." *The Nation*, May 23: 22–27.

Katz, Peter. 1993. *The New Urbanism: Toward an Architecture of Community*. McGraw-Hill.

Kevles, Daniel. 1997. *The Physicists: The History of a Scientific Community in Modern America*. Vintage.

Kim, Keechang. 2001. *Aliens in Medieval Law: The Origins of Modern Citizenship*. Cambridge University Press.

Kleiman, Jordan. 2000. The Appropriate Technology Movement in American Political Culture. Ph.D. dissertation, University of Rochester.

Kleinman, Daniel. 1995. *Politics on the Endless Frontier: Postwar Research Policy in the U.S.* Duke University Press.

Kleinman, Daniel. 2003. *Impure Cultures: University Biology at the Millennium*. University of Wisconsin Press.

Kleinman, Daniel, and Vallas, Steven. 2001. "Science, Capitalism, and the Rise of the 'Knowledge Worker': The Changing Structure of Knowledge Production in the United States." *Theory and Society* 30: 451–492.

Kline, Benjamin. 2000. *First Along the River: A Brief History of the U.S. Environmental Movement*. Acada Books.

Klonsky, Karen. 2000. "Forces Impacting the Production of Organic Foods." *Agriculture and Human Values* 17 (3): 233–243.

Knaap, Gerrit, and Emily Talen. 2005. "New Urbanism and Smart Growth: A Few Words from the Academy." *International Regional Science Review* 28 (2): 107–118.

Knorr-Cetina, Karin. 1981. *The Manufacture of Knowledge: An Essay on the Constructivist and Contextual Nature of Science.* Pergamon.

Kohler, Robert. 1991. *Partners in Science: Foundations and Natural Scientists, 1900–1945.* University of Chicago Press.

Korten, David. 1995. *When Corporations Rule the World.* Kumarian.

Korten, David. 1999. *The Post-Corporate World: Life after Capitalism.* Kumarian.

Krieger, Alex. 1998. "Whose Urbanism?" *Architecture* November, n.p. Retrieved September 1, 2005 from http://www.chankrieger.com.

Krimsky, Sheldon. 1982. *Genetic Alchemy: The Social History of the Recombinant DNA Controversy.* MIT Press.

Krimsky, Sheldon. 1992. "Regulating Recombinant DNA Research and Its Applications." In *Controversy*, third edition, ed. Dorothy Nelkin. Sage.

Kucinich, Dennis. 2004. "As Cleveland Mayor, Kucinich's Fight to Save Public Power." Retrieved March 31, 2006 from http://www.kucinich.us.

Kuhn, Thomas. 1970. *The Structure of Scientific Revolutions.* University of Chicago Press.

Kunstler, James. 1996. *Home from Nowhere: Remaking our Everyday World for the Twenty-First Century.* Simon and Schuster.

Lach, Jennifer. 2000. "Welcome to the Hoard Fest." *American Demographics* 22 (4): 8–9.

Lakatos, Imre. 1978. *The Methodology of Scientific Research Programmes.* Cambridge University Press.

Langevin, Mark, and Peter Rosset. 1997. "Land Reform from Below." *Food First Backgrounder* 4 (3): 1–4.

Latour, Bruno. 1987. *Science in Action: How to Follow Scientists and Engineers through Society.* Harvard University Press.

Latour, Bruno, and Steve Woolgar. 1986. *Laboratory Life: The Social Construction of Scientific Facts.* Princeton University Press.

Lawson, Laura. 2005. *City Bountiful: A Century of Community Gardening in the United States.* University of California Press.

Lennon, Barry, and David Richardson. 2002. "Teaching Old Dogs New Tricks: The Commercialization of Credit Unions." In *The Commercialization of Microfinance*, ed. Deborah Drake and Elisabeth Rhyne. Kumarian.

Leslie, Stuart. 1993. *The Cold War and American Science: The Military-Industrial-Academic Complex at MIT and Stanford.* Columbia University Press.

Levin, Hal. 2005. "Ten Basic Concepts for Architects and other Building Designer: Best Sustainable Indoor Air Quality Practices in Commercial Buildings." Retrieved March 31, 2006 from http://www.buildinggreen.com.

Levin, Morten. 1999. "Action Research Paradigms." In *Action Research*, ed. Davydd Greenwod. John Benjamins.

Levinger, David. 2002. Pedestrian Technologies: Redesigning Citizens, Organizers, and Technical Professionals. Ph.D. dissertation, Rensselaer Polytechnic Institute.

Liebowitz, Stan, and Stephen Margolis. 1990. "The Fable of the Keys." *Journal of Law and Economics* 33 (1): 1–25.

Lin, Biing-Hwan, Elizabeth Frazao, and Joanne Guthrie. 1999. "Away-from-Home Foods Increasingly Important to Quality of American Diet." *Agricultural Information Bulletin* No. 749, U.S. Department of Agriculture and U.S. Department of Health and Human Services. Retrieved March 31, 2006 from http://www.ers.usda.gov.

Linzey, Thomas, and Richard Grossman. 2004. "Shifting into a Different Gear: Empowering Communities, Protecting the Environment, and Building Democracy by Asserting Local Control over Factory Farm and Sludge Corporations in Pennsylvania." Retrieved March 31, 2006 from http://www.ratical.org.

Lipson, Mark. 1997. *Searching for the "O-Word": Analyzing the USDA Current Research Information System for Pertinence to Organic Farming. Executive Summary.* Retrieved March 31, 2006 from http://www.ofrf.org.

Little, Amanda. 2005. "Green vs. Green." *Wired*, February: 82.

Lockie, Stewart, Kristen Lyons, and Geoffrey Lawrence. 2000. "Constructing 'Green' Foods: Corporate Capital, Risk, and Organic Farming in Australia and New Zealand." *Agriculture and Human Values* 17 (4): 315–322.

Logan, John, and Harvey Molotch. 1987. *Urban Fortunes: The Political Economy of Place*. University of California Press.

Loka Institute. 2004a. "About the CRN." Retrieved March 31, 2006 from http://www.loka.org.

Loka Institute. 2004b. "Citizen Panels." Retrieved March 31, 2006 from http://www.loka.org.

Lotter, Don. 2000. "Brief History of Organic Agriculture and Overview of Its Major Institutions." Retrieved March 31, 2006 from http://entomology.ucdavis.edu.

Lounsbury, Michael, Marc Ventresca, and Paul M. Hirsch. 2003. "Social Movements, Field Frames, and Industry Emergence: A Cultural-Political Perspective on U.S. Recycling." *Socio-Economic Review* 1 (1): 71–105.

Lowi, Theodore. 1964. "American Business, Public Policy, Case-Studies, and Political Theory." *World Politics* 16 (4): 677–715.

Lowi, Theodore. 1972. "Four Systems of Policy, Politics, and Choice." *Public Administration Review* 32 (July-August): 298–310.

Luke, Timothy. 1997. *Ecocritique: Contesting the Politics of Nature, Economy, and Culture*. University of Minnesota Press.

Luke, Timothy. 1999. *Capitalism, Democracy, and Ecology: Departing from Marx*. University of Illinois Press.

Lukes, Steven. 1974. *Power: A Radical View*. Macmillan.

Lukes, Steven, ed. 1986. *Power*. New York University Press.

Lupo, Alan, Frank Colcord, and Edmund Fowler. 1971. *Rites of Way: The Politics of Transportation in Boston and the U.S. City*. Little, Brown.

MacKenzie, Donald, and Judy Wajcman. 1999. *The Social Shaping of Technology*. Open University Press

Madsen, Karen. 2002. "Green Havens." *E: the Environmental Magazine* 13 (5): 16–17.

Magill, John. 1994. "Credit Unions: A Formal-Sector Alternative for Financing Microenterprise Development." In *The New World of Microenterprise Finance*, ed. María Otero and Elisabeth Rhyne. Kumarian.

Maher, Joe. 2003. "Johnstown May Limit Yard Sales." *Schenectady Gazette*, June 28.

Marginson, Simon, and Mark Considine. 2000. *The Enterprise University: Power, Governance, and Reinvention in Australia*. Cambridge University Press.

Marshall, Alex. 2000. "Manifest Destiny: Is New Urbanism Just a Clever Way to Delay Real Solutions to Growth?" *Builder* 24 (3): 87–88.

Martin, Brian. 1988. "Nuclear Winter: Science and Politics." *Science and Public Policy* 15 (5): 321–334.

Martin, Brian. 1996. *Confronting the Experts*. SUNY Press.

Martin, Brian. 1999. "Suppression of Dissent in Science." In *Research in Social Problems and Public Policy*, Volume 7, ed. William R. Freudenburg and Ted I. K. Young. JAI.

Martin, Brian, ed. 1999. *Technology and Public Participation*, ed. Brian Martin. Retrieved March 31, 2006 from http://www.uow.edu.

Martin, Brian. 2006. "Publications on Whistleblowing and Suppression of Dissent." Retrieved March 31, 2006 from http://www.uow.edu.

Marx, Karl. 1977. *Capital*, volume 1. Random House.

Marx, Karl, and Frederich Engels. 1973. *The German Ideology*, part one, ed. C. J. Arthur. International Publishers.

Massachusetts Technology Collaborative. 2006. "The Big Picture." Retrieved March 31, 2006 from http://www.masstech.org.

Maurer, Bill. 2005. *Mutual Life, Limited: Islamic Banking, Alternative Currencies, Lateral Reason*. Princeton University Press.

Mauss, Marcel. 1967. *The Gift: Forms and Functions of Exchange in Archaic Societies*. Norton.

Mayer, Mayer, Eric Blank, and Blair Swezey. 1999. "The Grassroots Are Greener: A Community-Based Approach to Marketing Green Power." Retrieved January 20, 2005 from http://www.crest.org.

Mayntz, Renate, and Thomas Hughes, eds. 1988. *The Development of Large Technical Systems*. Campus Verlag.

Mazmanian, Daniel, and Michael Kraft, eds. 1999. *Toward Sustainable Communities: Transitions and Transformations in Environmental Policy*. MIT Press.

McAdam, Doug, and David Snow, eds. 1997. *Social Movements: Readings on Their Emergence, Mobilization, and Dynamics*. Roxbury.

McAdam, Doug, Sidney Tarrow, and Charles Tilly. 2001. *Dynamics of Contention*. Cambridge University Press.

McCally, Michael. 2002. "Social Policy and Social Movements: Medical Activism and Environmental Health." *Annals of the American Academy of Political and Social Science* 584: 145–158.

McCamant, Kathryn, and Charles Durrett with Ellen Hertzman. 1994. *Cohousing: A Contemporary Approach to Housing Ourselves*. Ten-Speed.

McDonough, William, and Michael Braungart. 2002. *Cradle-to-Cradle: Remaking the Way We Make Things*. North Point.

McFadden, Steven. 2004a. "The History of Community Supported Agriculture, Part I. Community Farms in the 21st Century: Poised for Another Wave of Growth?" Retrieved March 31, 2006 from http://www.newfarm.org.

McFadden, Steven. 2004b. "The History of Community Supported Agriculture, Part II. CSA's World of Possibilities." Retrieved March 31, 2006 from http://www.newfarm.org.

McGranahan, Gordon, and David Satterthwaite. 2000. "Environmental Health or Ecological Sustainability? Reconciling the Brown and Green Agendas in Urban Development." In *Sustainable Cities in Developing Countries*, ed. Cedric Pugh. Earthscan.

McNeely, Joseph. 1999. "Comment." In *Urban Problems and Community Development*, ed. Ronald Ferguson and William Dickens. Brookings Institution.

Melosi, Martin. 1981. *Garbage in the Cities: Refuse, Reform, and Environment, 1880–1890*. Texas A&M University Press.

Melucci, Alberto. 1980. "The New Social Movements: A Theoretical Approach." *Social Science Information* 52: 781–816.

Melucci, Alberto. 1996. *Challenging Codes: Collective Action in the Information Age*. Cambridge University Press.

Menard, Henry. 1971. *Science: Growth and Change*. Cambridge University Press.

Merton, Robert. 1973. *The Sociology of Science: Theoretical and Empirical Investigations*. University of Chicago Press.

Michels, Robert. 1958. *Political Parties: A Sociological Study of the Oligarchical Tendencies of Modern Democracy*. Free Press.

Micklethwait, John, and Adrian Wooldridge. 2003. *The Company: A Short History of a Revolutionary Idea*. Modern Library.

Microcredit Summit Campaign. 2006. "Microcredit Summit Campaign." Retrieved March 31, 2006 from http://www.microcreditsummit.org.

Mieszkowski, Katherine. 2004. "How George Bush Lost the Sun." Retrieved March 31, 2006 from http://www.truthout.org.

Miller, Nick. 1999. "Consignment Stores Set Standard." *Cincinnati Post*, online edition, May 18. Retrieved March 31, 2006 from http://www.cincypost.com.

Mitchell, Stacy. 2001. "Homegrown Economics: How Boulder Businesses Are Staying Ahead of the Chains." *Orion Afield*. Retrieved March 31, 2006 from http://www.newrules.org.

Mitchell, Stacy. 2003. "Independent Businesses, Unite!" *In Business*, July-August: 16–18. Retrieved March 31, 2006 from http://www.ilsr.org.

Mohl, Raymond. 2004. "Stop the Road: Freeway Revolts in American Cities." *Journal of Urban History* 30 (5): 674–706.

Mol, Arthur. 1995. *The Refinement of Production: Ecological Modernization Theory and the Chemical Industry*. International Books.

Mol, Arthur. 1996. "Ecological Modernization and Institutional Reflexivity: Environmental Reform in the Late Modern Age." *Environmental Politics* 5 (2): 302–323.

Mol, Arthur. 2000a. "The Environmental Movement in an Era of Ecological Modernization." *Geoforum* 31 (1): 45–56.

Mol, Arthur. 2000b. "Globalization and Environment: Between Apocalypse-Blindness and Ecological Modernization." In *Environment and Global Modernity*, ed. Gert Spaargaren, Arthur Mol, and Frederick Buttel. Sage.

Mol, Arthur. 2003. "The Environmental Transformation of the Modern Order." In *Modernity and Technology*, ed. Thomas Misa, Philip Brey, and Andrew Feenberg. MIT Press.

Mol, Arthur, and Gert Spaargaren. 2000. "Ecological Modernisation Theory in Debate: A Review." *Environmental Politics* 9 (1): 17–49.

Monahan, Torin. 2005. *Globalization, Technological Change, and Public Education*. Routledge.

Moore, Kelly. 1996. "Organizing Integrity: American Science and the Creation of Public Interest Science Organizations, 1955–1975." *American Journal of Sociology* 101 (6): 1592–1627.

Moore, Kelly. 2006a. *Doing Good While Doing Science*. Princeton University Press.

Moore, Kelly. 2006b. "Powered by the People: Scientific Authority in Participatory Science." In *The New Political Sociology of Science*, ed. Scott Frickel and Kelly Moore. University of Wisconsin Press.

Moore, Kelly, and Nicole Hala. 2002. "Organizing Identity: The Creation of Science for the People." *Research in the Sociology of Organizations* 19: 309–335.

Moore, Steven. 2007. "Architecture, Esthetics, and Public Health." Forthcoming in *The Difficult Dialogue*, ed. Sanda Iliescu. University of Virginia Press. Retrieved March 31, 2006 from http://www.ar.utexas.edu.

Moore, Steven, and Nathan Engstrom. 2005. "The Social Construction of 'Green Building' Codes: Competing Models by Industry, Government, and NGOs." In *Sustainable Architectures*, ed. Simon Guy and Steven A. Moore. Spoon.

Morgan, Sandra. 2002. *Into Our Own Hands: The Women's Health Movement in the United States, 1969–1990.* Rutgers University Press.

Morris, David. 2001. *Seeing the Light: Regaining Control of our Electricity System.* Institute for Local Self Reliance.

Morris, David. 2006. "Ownership Matters: Three Steps to Ensure a Biofuels Industry that Truly Benefits Rural America." Retrieved March 31, 2006 from http://www.newrules.org.

Moss, Ralph. 1996. *The Cancer Industry: The Classic Exposé on the Cancer Establishment.* Equinox.

Motavelli, Jim. 2001. "Zero Waste." *E Magazine* 12 (2). Retrieved March 31, 2006 from http://www.emagazine.com.

Moyer, Bill, with JoAnn McAllister, Mary Lou Finley, and Steven Soifer. 2001. *Doing Democracy: The MAP Model for Organizing Social Movements.* New Society.

Moyers, Bill. 2002. "Now with Bill Moyers: Trading Democracy." Public Broadcasting Corporation. Transcript retrieved March 31, 2006 from http://www.pbs.org.

Mullins, Nicholas. 1972. "The Development of a Scientific Specialty." *Minerva* 10: 51–82.

Mumford, Lewis. 1961. *The City in History: Its Origins, Its Transformations, and Its Prospects.* Harcourt Brace Jovanovich.

Mumford, Lewis. 1963. *Technics and Civilization.* Harcourt, Brace and World.

Murphy, Catherine. 1999. *Report No. 12: Cultivating Havana: Urban Agriculture and Food Security in the Years of Crisis.* Institute for Food and Development.

National Alliance of Clean Energy Business Incubators. 2002. "Clean Energy Alliance." Retrieved March 31, 2006 from http://www.incubator.com.

National Association of Home Builders. 2005. "NAHB's Model Green Home Building Guidelines." Retrieved March 31, 2006 from http://www.nahb.org.

National Association of Housing Cooperatives. 2006. "About NAHC and Housing Coops." Retrieved March 31, 2006 from http://www.coophousing.org.

National Association of Resale and Thrift Shops. 2006. "Press Kit." Retrieved March 31, 2006 from http://www.narts.org.

National Center for Appropriate Technology. 2003. "Fuel Funds and LIHEAP." Retrieved March 31, 2006 from http://www.ncat.org.

National Center for Appropriate Technology. 2006. "About NCAT: History." Retrieved March 31, 2006 from http://www.ncat.org.

National Community Capital Association. 2005. "Comparison of Types." Retrieved March 31, 2006 from http://www.communitycapital.org.

National Congress for Community Economic Development. 2001. "Policy on Faith-based and Community Development and Related Issues. Retrieved March 31, 2006 from http://www.ncced.org.

National Cooperative Grocers Association. 2002. "About Us: Roots of the NCGA." Retrieved March 31, 2006 from http://www.ncga.coop.

National Credit Union Administration. 2006a. "History of Credit Unions." Retrieved March 31, 2006 from http://www.ncua.gov.

National Credit Union Administration. 2006b. "NCUA History and Leadership." Retrieved March 31, 2006 from http://www.ncua.gov.

National Federation of Community Development Credit Unions. 2006a. "About Us." Retrieved July 5, 2003 from http://www.natfed.org.

National Federation of Community Development Credit Unions. 2006b. "Our History." Retrieved March 31, 2006 from http://www.natfed.org.

National Fuel Funds Network. 2006a. "Advocacy Tools." Retrieved March 31, 2006 from http://www.nationalfuelfunds.org.

National Fuel Funds Network. 2006b. "Who We Are." Retrieved March 31, 2006 from http://www.nationalfuelfunds.org.

National Grid. 2006. "GreenUp Providers." Retrieved March 31, 2006 from http://www.nationalgridus.com.

National Low Income Energy Consortium. 2006. "What Is NLIEC?" Retrieved March 31, 2006 from http://www.nliec.org.

NativeEnergy. 2006. "The Wind Farms." Retrieved March 31, 2006 from http://www.nativeenergy.com.

Natural Resources Defense Council. 1998. "Exhausted by Diesel: How America's Dependence on Diesel Engines Threatens Our Health." Retrieved March 31, 2006 from http://www.nrdc.org.

Natural Resources Defense Council. 2003. "Dirty Skies: The Bush Administration's Air Pollution Plan." Retrieved March 31, 2006 from http://www.nrdc.org.

Newton, David. 1996. *Environmental Justice: A Reference Handbook*. ABC-CLIO.

Noble, David. 1977. *America by Design: Science, Technology, and the Rise of Corporate Capitalism*. Oxford University Press.

Norman, Al. 2006a. "About Sprawl-Busters." Retrieved March 31, 2006 from http://www.sprawl-busters.com.

Norman, Al. 2006b. "Victorious Secret." Retrieved March 31, 2006 from http://www.sprawl-busters.com.

North, Peter. 2003. "Time Banks: Learning the Lessons from LETS?" *Local Economy* 18 (3): 267–270.

Northbourne, Walter. 1940. *Look to the Land*. Dent.

Northeast Organic Farming Association of New York. 2006. "New York Grown Chinese Medicinal Herbs Project." Retrieved March 31, 2006 from http://nofany.org.

Northeast Organic Farming Association of Vermont. 2006. "About NOFA Vermont!" Retrieved January 5, 2004 from http://www.nofavt.org.

Nowotny, Helga, and Rose, Hilary, eds. 1979. *Counter-Movements in the Sciences: the Sociology of the Alternatives to Big Science*. Reidel.

Nowotny, Helga, Peter Scott, and Michael Gibbons, eds. 2001. *Re-Thinking Science: Knowledge and the Public in an Age of Uncertainty*. Oxford University Press.

Nutrition Business Journal. 2001. "The U.S. Organic Industry III." *Nutrition Business Journal* 6 (2): 1–10.

Nutrition Business Journal. 2004. "NBJ's Organic Foods Report 2004." Retrieved March 31, 2006 from http://store.yahoo.com.

Nye, David. 1998. *Consuming Power: A Social History of American Energies*. MIT Press.

Obach, Brian. 2002. "Labor-environmental Relations: An Analysis of the Relations between Labor Unions and Environmentalists." *Social Science Quarterly* 83 (1): 82–100.

O'Brien, Robert, Anne Goetz, Jan Scholter, and Marc Williams. 2000. *Contesting Global Governance: Multilateral Economic Institutions and Global Social Movements*. Cambridge University Press.

O'Connor, Alice. 1999. "Swimming against the Tide." In *Urban Problems and Community Development*, ed. Ronald Ferguson and William Dickens. Brookings Institution.

O'Connor, James. 1998. *Natural Causes? Essays in Ecological Marxism*. Guilford.

O'Keefe, Karen. 2002. "Peter Katz: Marketing the New Urbanism." *The Town Paper* 4 (4): 1–3. Retrieved March 31, 2006 from http://www.tndtownpaper .com.

Organic Consumers Association. 2006. "About the OCA." Retrieved March 31, 2006 from http://www.organicconsumers.org.

Organic Farming Research Foundation. 2003. *State of the States, 2nd Edition: Organic Systems Research at Land Grant Institutions, 2001– 2003*. Retrieved March 31, 2006 from http://www.ofrf.org.

Organic Trade Association. 2004. "2004 Organic Cotton Survey." Retrieved March 31, 2006 from http://www.ota.com.

Organic Trade Association. 2005. "Organic Fashion Comes of Age." Retrieved March 31, 2006 from http://www.ota.com.

Organic Trade Association. 2006. "OTA Accomplishments." Retrieved March 31, 2006 from http://www.ota.com.

Organic Valley Family of Farms. 2006. "Our History: Born in a Barn!" Retrieved March 31, 2006 from http://organicvalley.coop.

O'Rourke, Dara. 2005. "Market Movements: Nongovernmental Organization Strategies and Consumption." *Journal of Industrial Ecology* 9 (1-2): 115–128.

Palmer, Karen, and Margaret Walls. 2002. "The Product Stewardship Movement: Understanding Costs, Effectiveness, and the Role of Policy." *Resources for the Future Report*, November. Retrieved March 31, 2006 from http://www.rff.org.

PARnet. 2006. "Welcome to PARnet." Retrieved March 31, 2006 from http://www.parnet.org.

Parsons, Talcott. 1951. *The Social System*. Free Press.

Parsons, Talcott. 1977. *The Evolution of Societies*. Prentice-Hall.

Pauli, Gunter. 1998. *The Road to Zero Emissions: More Jobs, More Income, and No Pollution*. Greenleaf.

Pellow, David. 2002. *Garbage Wars: The Struggle for Environmental Justice in Chicago*. MIT Press.

Pellow, David. 2005. "Transnational Alliances and Global Politics: New Geographies of Urban Environmental Justice Struggles." Paper presented at annual meeting of American Sociological Association, Philadelphia.

Pellow, David, and Robert Brulle, eds. 2005. *Power, Justice, and the Environment: A Critical Appraisal of the Environmental Justice Movement*. MIT Press.

Pellow, David, and Lisa Sun-Hee Park, 2002. *Silicon Valley of Dreams: Environmental Injustice, Immigrant Workers, and the High-Tech Global Economy*. New York University Press.

Pellow, David, Allan Schnaiberg, and Adam Weinberg. 2000. "Putting the Ecological Modernization Theory to the Test: The Promises and Performances of Urban Recycling." *Environmental Politics* 9: 109–137.

Perkins, John. 2004. *Confessions of an Economic Hit Man*. Berrett-Koehler.

Pesticide Action Network North America. 2003. "About Us." Retrieved March 31, 2006 from http://www.panna.org.

Peters, Suzanne. 1979. "Organic Farmers Celebrate Organic Research: A Sociology of Popular Science." In *Counter-Movements in the Sciences*, ed. Helga Nowotny and Hilgary Rose. Reidel.

Pichardo, Nelson. 1997. "New Social Movements: A Critical Review." *Annual Review of Sociology* 23: 411–430.

Pinch, Trevor, and Wiebe Bijker. 1999. "The Social Construction of Facts and Artifacts: Or How the Sociology of Science and the Sociology of Technology Might Benefit Each Other." In *The Social Construction of Technological Systems*, ed. Wiebe Bijker, Thomas Hughes, and Trevor Pinch. MIT Press.

Pinderhughes, Raquel. 2004. *Alternative Urban Futures: Planning for Sustainable Development in Cities throughout the World*. Rowman and Littlefield.

Piven, Frances, and Richard Cloward. 1971. *Regulating the Poor: The Functions of Public Welfare*. Pantheon.

Piven, Frances, and Richard Cloward. 1977. *Poor People's Movements: Why They Succeed, How They Fail*. Pantheon.

Podobnik, Bruce. 2004. "Resistance to Globalization: Cycles and Evolutions in the Globalization Protest Movement." Presented at annual meeting of American Sociological Association, San Francisco.

Podobnik, Bruce, and Thomas Reifer. 2004. "The Globalization Protest Movement in Comparative Perspective." *Journal of World Systems Research* 10 (1): 3–9.

Pollack, Roger. 1984. "Solar Power: The Promise Fades." *The Progressive* 48 (9): 32–34.

Pollan, Michael. 2001. "Naturally." *New York Times Magazine*, May 13: 30ff.

Ponte, Stefano. 2004. "Standards and Sustainability in the Coffee Sector: A Global Value Chain Approach." Retrieved March 31, 2006 from http://www.iisd.org.

Poppendieck, Janet. 1998. *Sweet Charity? Emergency Food and the End of Entitlement.* Viking.

Porter, Michael, and Claas van der Linde. 1995. "Green and Competitive." *Harvard Business Review* 68 (3): 79–91.

Portney, Kent. 2002. *Taking Sustainability Seriously: Economic Development, the Environment, and Quality of Life in American Cities.* MIT Press.

Pozdena, Randall. 2002. "Smart Growth and Its Effects on Housing Markets: The New Segregation." Retrieved March 31, 2006 from http://www.nationalcenter.org.

Price, Derek de Solla. 1963. *Little Science, Big Science.* Columbia University Press.

Price, David. 1989. *Before the Bulldozer: The Nambiquara Indians and the World Bank.* Seven Locks.

Princen, Thomas, Michael Maniates, and Ken Conca, eds. 2002. *Confronting Consumption.* MIT Press.

Public Citizen. 2006a. "The Case for Renewables and Energy Efficiency." Retrieved March 31, 2006 from http://www.citizen.org.

Public Citizen. 2006b. "Stopping the Nuclear Relapse." Retrieved March 31, 2006 from http://www.citizen.org.

Puckett, John, and Ira Harkavy. 1999. "The Action Research Tradition in the United States." In *Action Research*, ed. Davydd Greenwod. John Benjamins.

Puffert, Douglas. 2003. "Path Dependence." *EH Net Encyclopedia*, ed. Robert Whaples. Retrieved March 31, 2006 from http://eh.net.

Purdue, Derrick. 2000. *Anti-GenetiX: The Emergence of the Anti-GM Movement.* Ashgate.

Pyatok, Michael. 2000. "The Politics of Design: The New Urbanists vs. the Grassroots." *Housing Policy Debate* 11 (4): 803–814.

Pynchon, Thomas. 1973. *Gravity's Rainbow.* Viking.

Quine, Willard Van Orman. 1980. "Two Dogmas of Empiricism." In *Challenges to Empiricism*, ed. Harold Morick. Hackett.

Rainforest Action Network. 1996. "Three Activists Scale Mitsubishi Building with Huge Banner in New York City." Retrieved March 31, 2006 from http://ran.org.

Rainforest Action Network. 1998. "Landmark Settlement Reached in Long-running Environmental Boycott of Two Mitsubishi Companies." Retrieved March 31, 2006 from http://ran.org.

Rainforest Action Network. 2002. "Mitsubishi Corporation Supports Forest Stewardship Council Certification." Retrieved March 31, 2006 from http://ran.org.

Rainforest Action Network. 2006a. "The Global Finance Campaign." Retrieved March 31, 2006 from http://ran.org.

Rainforest Action Network. 2006b. "What We Do: History." Retrieved March 31, 2006 from http://ran.org.

Rainforest Action Network. 2006c. "Jumpstart Ford." Retrieved March 31, 2006 from http://ran.org.

Raynolds, Laura. 2000. "Re-embedding Global Agriculture: The International Organic and Fair Trade Movements." *Agriculture and Human Values* 17: 297–309.

Raynolds, Laura. 2002. "Consumer/producer Links in Fair Trade Coffee Networks." *Sociologia Ruralis* 42 (4): 404–424.

Reardon, Thomas, and Elizabeth Farina. 2002. "The Rise of Private Food Quality and Safety Standards: Illustrations from Brazil." *International Food and Agricultural Management Review* 4: 413–421.

Reece, Ray. 1979. *The Sun Betrayed: A Report on the Corporate Seizure of U.S. Solar Energy Development*. South End.

Reed, Matthew. 2001. "Fight the Future! How the Contemporary Campaigns of the U.K. Organic Movement Have Arisen from Their Composting of the Past." *Sociologia Ruralis* 41 (1): 131–145.

Regional Commission on Airport Affairs. 2005. "Airport Community Groups and Cities." Retrieved March 31, 2006 from http://www.rcaanews.org.

Reisner, Ann Elizabeth. 2001. "Social Movement Organizations' Reactions to Genetic Engineering in Agriculture." *The American Behavioral Scientist* 44 (8): 1389–1404.

Renner, Michael. 2001. "U.S. Contempt for Alternatives." *World Watch* 14 (5): 2.

Rennings, Klaus. 2000. "Redefining Innovation: Eco-Innovation Research and the Contribution from Ecological Economics." *Ecological Economics* 32: 319–332.

Rescher, Nicholas. 1978. *Scientific Progress: A Philosphical Essay on the Economics of Research in Natural Science*. University of Pittsburgh Press.

Restivo, Sal. 1983. "The Myth of the Kuhnian Revolution." In *Sociological Theory*, ed. Randall Collins. Jossey-Bass.

Rhyne, Elisabeth. 2002. "Commercialization and Crisis in Microfinance." In

The Commercialization of Microfinance, ed. Deborah Drake and Elisabeth Rhyne. Kumarian.

Richards, Evelleen. 1981. *Vitamin C and Cancer: Medicine or Politics?* St. Martin's Press.

Righter, Robert. 1996. *Wind Energy in America: A History.* University of Oklahoma Press.

Ringer, Fritz. 2004. *Max Weber: An Intellectual Biography.* University of Chicago Press.

Ritscher, Adam. n.d. "Blood on the Scarecrow: The Ongoing Farm Crisis." Retrieved March 31, 2006 from http://www.lefthook.org.

Roberts, J. Timmons, and Melissa Toffolon-Weiss. 2001. *Chronicles from the Environmental Justice Frontline.* Cambridge University Press.

Robertson, Roland. 1995. "Glocalization: Time-Space and Homogeneity-Heterogeneity." In *Global Modernities*, ed. Mike Featherstone, Scott Lash, and Roland Robertson. Sage.

Robinson, Lynn. 2002. "Doing Good and Doing Well: Shareholder Activism, Responsible Investment, and Mainline Protestantism." In *The Quiet Hand of God*, ed. Robert Wuthnow and John Evans. University of California Press.

Rodale, Jerome. 1948. *The Organic Front.* Rodale.

Rodale Institute. 2006. "Who We Are: History." Retrieved March 31, 2006 from http://www.rodaleinstitute.org.

Roosens, Eugene. 1989. *Creating Ethnicity: The Process of Ethnogenesis.* Sage.

Rose, Fred. 2000. *Coalitions across the Class Divide: Lessons from the Labor, Peace, and Environmental Movements.* Cornell University Press.

Rosset, Peter. 2001. "Tides Shift on Agrarian Reform: New Movements Show the Way." *Food First Backgrounder* 7 (1): 1–8. Institute for Food and Development Studies.

Rossiter, Margaret. 1993. "The Matthew Matilda Effect in Science." *Social Studies of Science* 23: 324–341.

Rusk, Daivd. 1999. *Inside Game, Outside Game: Winning Strategies for Saving Urban America.* Brookings Institution Press.

Russell, Dorlene. 2000. "Save Your Landfill: Have a Garage Sale." *Waste Age* 31 (3): 18–19.

R. W. Beck, Inc. 2001. *Final Report: U.S. Recycling Information Study.* National Recycling Coalition, Inc. Retrieved March 31, 2006 from http://www.epa.gov.

Sahlins, Marshall. 1976. *Culture and Practical Reason.* University of Chicago Press.

Salamon, Lester, and Wojciech Sokolowski. 2004. *Global Civil Society: Dimensions of the Nonprofit Sector*, volume 2. Kumarian.

Salvation Army. 2002. "International Statistics." Retrieved March 31, 2006 from http://www1.salvationarmy.org.

Sassen, Saskia. 2000. *Cities in a World Economy.* Pine Forge/Sage.

Sassen, Saskia. 2003. "Economic Globalization and the Redrawing of Citizenship." In *Globalization, the State, and Violence*, ed. Jonathan Friedman. Altamira.

Saxenian, Annalee. 1996. *Regional Advantage: Culture and Competition in Silicon Valley and Route 128*. Harvard University Press.

Sazama, Gerald. 2000. "Lessons from the History of Affordable Housing Cooperatives in the United States: A Case Study in American Affordable Housing Policy." *American Journal of Economics and Sociology* 59 (4): 575–608.

Schäfer, Wold. 1983. *Finalization in Science: The Social Orientation of Scientific Progress*. Reidel.

Scheinberg, Anne. 2003. "The Proof of the Pudding: Urban Recycling in North America as a Process of Ecological Modernization." *Environmental Politics* 12 (4): 49–75.

Schlarb, Mary. 2001. "Eco-Industrial Development: A Strategy for Building Communities." *Review of Economic Development Literature and Practice No. 8*. U.S. Economic Development Administration, U.S. Department of Commerce.

Schlett, James. 2006. "Small Credit Unions Getting Consulting Help." *Schenectady Gazette* Feburary 26.

Schmitt, John. 2000. "Inequality and Globalization: Some Evidence from the United States." In *The Ends of Globalization*, ed. Don Kalb, Marco van der Lind, Richard Staring, Bart van Steenbergen, and Nico Wilterdink. Rowman and Littlefield.

Schnaiberg, Allan. 1982. "Did You Ever Meet a Payroll? Contradictions in the Structure of the Appropriate Technology Movement." *Humboldt Journal of Social Relations* 9 (2): 38–62.

Schnaiberg, Allan. 1983a. "Redistributive Goals versus Distributive Politics: Social Equity Limits in Environmental and Appropriate Technology Movements." *Sociological Inquiry* 53: 200–219.

Schnaiberg, Allan. 1983b. "Soft Tech/Hard Tech, Hi Tech/Lo Tech: A Social Movement Analysis of Appropriate Technology." *Sociological Inquiry* 53: 220–251.

Schnaiberg, Allan. 1983c. "Soft Energy and Hard Labor? Structural Restraints on the Transition to Appropriate Technology." In *Technology and Social Change in Rural Areas*, ed. Gene Summers. Westview.

Schnaiberg, Allan, and Kenneth Gould. 1994. *Environment and Society: The Enduring Conflict*. St. Martin's Press.

Schneider, Mindi. 2005. "Ownership and Control in the Organic Retail Sector: A Critical Examination." Paper presented at annual meeting of Agriculture, Food, and Human Values Society, Portland, Oregon.

Schumacher, E. F. 1973. *Small Is Beautiful: Economics as if People Mattered*. Harper and Row.

Schumpeter, Joseph. 1975. *Capitalism, Socialism, and Democracy*. Harper and Row.

Schurman, Rachel. 2004. "Fighting 'Frankenfoods': Industry Opportunity Structures and the Efficacy of the Anti-Biotech Movement in Western Europe." *Social Problems* 51 (2): 243–268.

Schwartz, Rachel. 2005. "An Analysis of Corporate Control in Food Retailing." Paper presented at annual meeting of Agriculture, Food, and Human Values Society, Portland, Oregon.

Sclove, Richard. 1995a. *Democracy and Technology*. Guilford.

Sclove, Richard. 1995b. "Putting Science to Work in Communities." *Chronicle of Higher Education* 41 (29): B1–B3.

Seavy, Mark. 2000. "Value Village, Savers Lead Thrift Store Growth." Retrieved March 31, 2006 from http://www.icsc.org.

Seldman, Neil. 1995. "Recycling: A History in the United States." In *Encyclopedia of Energy, Technology, and the Environment*, ed. Atillio Bisio and Sharon Boots. Wiley.

Sen, Amartya. 1999. *Development as Freedom*. Knopf.

Sevron, Lisa. 2002. "Fulfilling the Potential of the U.S. Microenterprise Strategy." In *Replicating Microfinance in the United States*, ed. James Carr and Zhong Yi Tong. Woodrow Wilson Center Press.

Seyfang, Gill. 2003. "'With a Little Help from my Friends': Evaluating Time Banks as a Tool for Community Self-help." *Local Economy* 18 (3): 257–264.

Seyfang, Gill. 2004. "Time Banks: Rewarding Self-Help in the Inner City?" *Community Development Journal* 39 (1): 62–71.

Shapin, Steven, and Simon Schaffer. 1985. *Leviathan and the Air Pump: Hobbes, Boyle, and the Experimental Life*. Princeton University Press.

Shellenberger, Michael, and Ted Nordhaus. 2004. "The Death of Environmentalism: Global Warming Politics in a Post-Environmental World." Retrieved March 31, 2006 from http://thebreakthrough.org.

Shorebank Pacific. 2004. "About Us." Retrieved March 31, 2006 from http://www.eco-bank.com.

Shorto, Russell. 2004. "A Short-Order Revolutionary." *New York Times Magazine*, January 11: 18–21.

Shuman, Michael. 2000. *Going Local: Creating Self-Reliant Communities in a Global Age*. Routledge.

Shuman, Michael. 2006. *The Small Mart Revolution: How Local Businesses Are Beating the Global Competition*. Berrett-Koehler.

Siegler, Richard, and Herbert Levy. n.d. "Brief History of Cooperative Housing." Retrieved March 31, 2006 from http://www.coophousing.org.

Sierra Club. 2006a. "Cheney Energy Task Force Details Remain Hidden." Retrieved March 31, 2006 from http://www.sierraclub.org.

Sierra Club. 2006b. "Stop Sprawl: Anti-sprawl Campaigns." Retrieved January 18, 2005 from http://www.sierraclub.org.

Simpson, Christopher, ed. 1998. *Universities and Empire: Money and Politics in the Social Sciences During the Cold War*. New Press.

Sissine, Fred. 1999. "Renewable Energy: Key to Sustainable Energy Supply." *Congressional Research Service Report to Congress 97031*. Library of Congress. Retrieved March 31, 2006 from http://www.ncseonline.org.

Sklair, Leslie. 2001. *The Transnational Capitalist Class*. Blackwell.

Slaughter, Sheila, and Larry Leslie. 1997. *Academic Capitalism: Politics, Policies, and the Entrepreneurial University*. Johns Hopkins University Press.

Slaughter, Sheila, and Larry Leslie. 2001. "Expanding and Elaborating the Concept of Academic Capitalism." *Organization* 8 (2): 154–161.

Smart Growth America. 2006. "Smart Growth America." Retrieved March 31, 2006 from http://www.smartgrowthamerica.com.

Smart Growth Network. 2003. "What It Is; What It Does." Retrieved March 31, 2006 from http://www.smartgrowth.org.

Smith, Everard, and Terry Marsden. 2003. "Exploring the 'Limits to Growth' in UK Organics: Beyond the Statistical Image." Working Paper Series No. 3, Centre for Business Relationships, Accountability, Sustainability, and Society, Cardiff University. Retrieved March 31, 2006 from http://www.brass.cf.ac.uk.

Smith, Jackie. 2004. "Exploring Connections between Global Integration and Political Mobilization." *Journal of World-Systems Research* 10 (1): 255–285.

Snow, David. 2002. "Social Movements as Challenges to Authority: Resistance to an Emerging Conceptual Hegemony." Presented at American Sociological Association workshop, University of Notre Dame.

Social Investment Forum. 2001. "2001 Report on Socially Responsible Investing Trends in the United States." Social Investment Forum. Retrieved March 31, 2006 from http://www.socialinvest.org.

Social Investment Forum. 2005. "Socially Responsible Mutual Funds." Retrieved March 31, 2006 from http://www.socialinvest.org.

Solar Region Freiburg. 2003. "Freiburg Solar City." Retrieved March 31, 2006 from http://www.solarregion.freiburg.de.

Spaargaren, Gert, and Bas van Vliet. 2000. "Lifestyles, Consumption, and the Environment: The Ecological Modernization of Domestic Consumption." *Environmental Politics* 9 (1): 50–77.

Spencer, Marty Traynor, and Patrick Rhea. 2003. "Market Overview: Sales Top $36B." Natural Foods Merchandiser, June online edition. Retrieved June 24, 2003 from http://exchange.healthwell.com.

Spiegelman, Helen. 2003. "Beyond Recycling: The Future of Waste." Grassroots Recycling Organization. Retrieved March 31, 2006 from http://www.grrn.org.

Star, Susan Leigh, and James Greisemer. 1989. "Institutional Ecologies, 'Translations,' and Boundary Objects: Amateurs and Professionals in Berkeley's Museum of Vertebrate Zoology, 1907–39." *Social Studies of Science* 19: 387–420.

Stephenson, Bruce. 2002. "The Roots of New Urbanism." *Journal of Planning History* 1 (2): 99–123.

Stevenson, Steve, Daniel Lass, John Hendrickson, and Kathy Ruhf. 2004. "CSA across the Nation: Findings from 1999 and 2001 National Surveys." Presented at annual meeting of Agricultural, Food, and Human Values Society, June, Hyde Park, N.Y. An earlier report was retrieved March 31, 2006 from http://www.cias.wisc.edu.

Stoecker, Randy. 1997. "The CDC Model of Urban Redevelopment: A Critique and an Alternative." *Journal of Urban Affairs* 19 (1): 1–22.

Stoutland, Sara. 1999. "Community Development Corporations: Mission, Strategy, and Accomplishment." In *Urban Problems and Community Development*, ed. Ronald Ferguson and William Dickens. Brookings Institution.

Strand, Kerry, Sam Marullo, Nick Cutforth, Randy Stoecker, and Patrick Donahue. 2003. *Community-Based Research and Higher Education: Principles and Practices*. Jossey-Bass.

Strasser, Susan. 1999. *Waste and Want: A Social History of Trash*. Metropolitan Books, Henry Holt and Co.

Strathern, Marilyn, ed. 2000. *Audit Cultures: Anthropological Studies in Accountability, Ethics, and the Academy*. Routledge.

Strohm, Chris. 2003. "How Seven Families Saved Their Farms by Going Organic." *Co-op America Quarterly* 60 (summer): 13.

Summit II National Office. 2002a. "Summit II Briefing: Celebrating Our Victories, Strengthening Out Roots." Retrieved March 31, 2006 from http://www.ejrc.cau.edu.

Summit II National Office. 2002b. "Environmental Justice Timeline: Milestones." Retrieved March 31, 2006 from http://www.ejrc.cau.edu.

Sunnucks. Mike. 2001. "Anti-sprawl Groups Band Together to Oppose Techway." *Washington Business Journal* March 30, n.p. Retrieved March 31, 2006 from http://www.bizjournals.com.

Swanson, Walden, Peg Nolan, and Dave Gutknecht. 2001. "Retail Operations Survey 2000." *Cooperative Grocer Online* 95, July-August, n.p. Retrieved March 31, 2006 from http://www.cooperativegrocer.coop.

Swartz, David. 1997. *Culture and Power: The Sociology of Pierre Bourdieu*. University of Chicago Press.

Szasz, Andrew. 1994. *Ecopopulism: Toxic Waste and the Movement for Environmental Justice*. University of Minnesota Press.

Talen, Emily. 2005. *New Urbanism and American Planning: The Conflict of Cultures*. Routledge.

Tanaka, Keiko, and Lawrence Busch. 2003. "Standardization as a Means for Globalizing a Commodity: The Case of Rapeseed in China." *Rural Sociology* 68 (1): 25–45.

Tarasuk, Valerie, and Joan Eakin. 2003. "Charitable Food Assistance as Symbolic Gesture: An Ethnographic Study of Food Banks in Ontario." *Social Science and Medicine* 56 (7): 1505–1515.

Tarrow, Sidney. 1998. *Power in Movement: Social Movements and Contentious Politics*. Cambridge University Press.

Tatum, Jesse. 1994. "Technology and Values: Getting Beyond the 'Device Paradigm' Impasse." *Science, Technology, and Human Values* 19 (1): 70–87.

Tatum, Jesse. 1995. *Energy Possibilities: Rethinking Alternatives and the Choice-Making Process*. SUNY Press.

Tatum, Jesse. 2000. *Muted Voices: The Recovery of Democracy in the Shaping of Technology*. Associated University Presses.

Taylor, Dorceta. 2000. "The Rise of the Environmental Justice Paradigm: Injustice Framing and the Social Construction of Environmental Discourses." *American Behavioral Scientist* 43 (4): 508–580.

Thomsen, Mark. 2001. "Environmental Groups Endorse Shareholder Resolutions." Institutional Shareowner. Retrieved March 31, 2006 from http://www.institutionalshareowner.com.

Thornton, Joe. 2000. *Pandora's Poison: Chlorine, Health, and a New Environmental Strategy*. MIT Press.

Time Dollar USA. 2006a. "Mission." Retrieved March 31, 2006 from http://www.timedollar.org.

Time Dollar USA. 2006b. "Our Support." Retrieved March 31, 2006 from http://www.timedollar.org.

Todd, Nancy, and John Todd. 1993. *From Eco-Cities to Living Machine*. North Atlantic Books.

Touraine, Alain. 1992. "Beyond Social Movements?" *Theory, Culture, and Society* 9 (1): 125–145.

Transfair USA. 2006. "History." Retrieved March 31, 2006 from http://www.transfairusa.org.

Traweek, Sharon. 1988. *Beamtimes and Lifetimes: The World of High-Energy Physicists*. Harvard University Press.

Traweek, Sharon. 1992. "Border Crossings: Narrative Strategies in Science Studies and among Physicists in Tsukuba Science City, Japan." In *Science as Practice and Culture*, ed. Andrew Pickering. University of Chicago Press.

Truffer, Bernard, and Gregor Dürrenberger. 1997. "Outsider Initiatives in the Reconstruction of the Car: The Case of Lightweight Vehicle Milieus in Switzerland." *Science, Technology, and Human Values* 22 (2): 207–234.

Union of Concerned Scientists. 2005. "Food and Environment." Retrieved March 31, 2006 from http://www.ucsusa.org.

Union of Concerned Scientists. 2006. "Clean Energy." Retrieved March 31, 2006 from http://www.ucsusa.org.

United Nations Development Program. 1996. *Urban Agriculture: Food, Jobs, and Sustainable Cities*. United Nations Development Programme Publication Series for Habitat II, Volume 1.

United Nations Development Program. 1999. *Human Development Report 1999*. Oxford University Press.

Urban Ecovillage Network. 2004. "About Us." Retrieved March 31, 2006 from http://urban.ecovillage.org.

U.S. Census Bureau. 1995. "Table 1. Urban and Rural Population: 1900 to 1990." Retrieved March 31, 2006 from http://www.census.gov.

U.S. Department of Agriculture. 2000. "Farmers' Markets Facts!" Retrieved March 31, 2006 from http://www.ams.usda.gov.

U.S. Department of Agriculture. 2004. "Farmers Market Growth." Retrieved March 31, 2006 from http://www.ams.usda.gov.

U.S. Department of Energy. 2003. "Status and Impact of State MTBE Ban." Energy Information Administration. Retrieved March 31, 2006 from http://www.eia.doe.gov.

U.S. Department of Energy. 2004. "Top Ten Utility Green Pricing Programs." Retrieved March 31, 2006 from http://www.eere.energy.gov.

U.S. Department of Energy. 2006. "Eliminating MTBE in Gasoline in 2006." Retrieved March 31, 2006 from http://www.eia.doe.gov.

U.S. Department of Health and Human Services. 1999. "Renewables Gaining Ground through State SWAP, Tribal Pilot Programs." LIHEAP Clearinghouse. Retrieved March 31, 2006 from http://www.ncat.org.

U.S. Environmental Protection Agency. 2004a. "Evaluation Report: EPA Needs to Consistently Implement the Intent of the Executive Order on Environmental Justice." Report No. 2004-P-00007, March. Retrieved March 31, 2006 from http://www.epa.gov.

U.S. Environmental Protection Agency2004b. "Science of Environmental Justice Working Conference." Retrieved March 31, 2006 from http://www.epa.gov.

U.S. Green Building Council. 2005. "USBCB Backgrounder." Retrieved March 31, 2006 from http://www.usgbc.org.

U.S. Green Building Council. 2006. "LEED Certification Process." Retrieved March 31, 2006 from http://www.usgbc.org.

Van der Ryn, Sim, and Peter Calthorpe. 1986. *Sustainable Communities: A New Design Synthesis for Cities, Suburbs, and Towns.* Sierra Club Books.

Van Til, Jan. 2000. *Growing Civil Society: From Non-Profit Sector to Third Space.* Indiana University Press.

Varma, Roli. 1995. "Restructuring Corporate Research and Development: From an Autonomous to a Linkage Model." *Technology Analysis and Strategic Management* 7 (2): 231–247.

Varma, Roli. 1999. "Professional Autonomy vs. Industrial Control?" *Science as Culture* 8 (1): 23–45.

Von Hassell, Malve. 2002. *The Struggle for Eden: Community Gardens in New York City.* Bergin and Garvey.

Vos, Timothy. 2000. "Visions of the Middle Landscape: Organic Farming and the Politics of Nature." *Agriculture and Human Values* 17 (3): 245–256.

Wallerstein, Immanuel. 1974. *The Modern World System*, volume 1: *Capitalist Agriculture and the Origins of the European World Economy in the Sixteenth Century*. Academic Press.

Walsh, Edward, Rex Warland, and D. Clayton Smith. 1997. *Don't Burn It Here: Grassroots Challenges to Trash Incinerators*. Pennsylvania State University Press.

Warren, Tim. 1999. "Onward Christian Soldiers." *Washingtonian* 35 (3): 86–89.

Warwick, Hugh. 2001. "Cuba's Organic Revolution." *Forum for Applied Research and Public Policy* 16 (2): 54–58.

Weber, Max. 1949. *The Methodology of the Social Sciences*. Free Press.

Weber, Max. 1978. *Economy and Society*. University of California Press.

Weeks, Electra. 2004. "Where Have all the Public Scientists Gone? Agricultural Biotechnology in South Indian and the Changing Role of Public Scientists and Public Science." Paper presented at the conference on Science, Technology, and the Environment, Rensselaer Polytechnic Institute.

Weinberg, Adam. 1998. "Distinguishing among Green Businesses: Growth, Green, and Anomie." *Society and Natural Resources* 11 (3): 241–250.

Weinberg, Adam, David Pellow, and Allan Schnaiberg. 2000. *Urban Recycling and the Search for Sustainable Community Development*. Princeton University Press.

Weintraub, Arlene, and Pallavi Gogoi. 2003. "The Outcry over 'Terminator' Genes." *Business Week*, July 14: 70–71.

Weir, David, and Mark Schapiro. 1981. *Circle of Poison: Pesticides and People in a Hungry World*. Institute for Food and Development Policy.

Weir, Margaret. 1999. "Power, Money, and Politics in Community Development." In *Urban Problems and Community Development*, ed. Ronald Ferguson and William Dickens. Brookings Institution.

Weisbrod, Burton. 1988. *The Nonprofit Economy*. Harvard University Press.

Weisbrod, Burton, ed. 1998. *To Profit or Not to Profit: The Commercial Transformation of the Nonprofit Sector*. Cambridge University Press.

Wellock, Thomas. 1998. *Critical Masses: Opposition to Nuclear Power in California, 1958–1978*. University of Wisconsin Press.

Welsh, Heidi. 1988. "Shareholder activism." *Multinational Monitor* 9 (12), n.p. Retrieved March 31, 2006 from http://multinationalmonitor.org.

White, Victoria, and Anita Campion. 2002. "Transformation: Journey from NGO to Regulated MFI." In *The Commercialization of Microfinance*, ed. Deborah Drake and Elisabeth Rhyne. Kumarian.

Whorisky, Peter. 2005. "Residents Find Small-Town Vision Blurred: In Clarksburg, Home Buyers Accuse Developer of Betraying Original Plan." *Washington Post*, July 31.

Wilcox, James. 2005. "Economies of Scale and Continuing Consolidation of Credit Unions." Federal Reserve Bank of San Francisco Economic Letter 2005-29, November 4. Retrieved March 30, 2006 from http://www.frbsf.org.

Williamson, Thad, David Imbroscio, and Gar Alperovitz. 2002. *Making a Place for Community: Local Democracy in a Global Era.* Routledge.

Wilson, S. S. 1973. "Bicycle Technology." *Scientific American*, March: 81–100.

Winne, Mark. 2006. "Meat-Industrial Complex: How Factory Farms Undercut Public Health." *In These Times*, March: 30–31.

Winner, Langdon. 1977. *Autonomous Technology: Technics Out-of-Control as a Theme in Political Thought.* University of California Press.

Winner, Langdon. 1986. *The Whale and the Reactor: The Search for Limits in an Age of High Technology.* University of Chicago Press.

Winner, Langdon, and David Hess. 2007. *Sustainable Technologies and Democratic Activism.* Forthcoming.

Wirka, Susan. 1996. "The City Social Movement: Progressive Women Reformers and Early Social Planning." In *Planning the Twentieth-Century American City*, ed. Mary Sies and Christopher Silver. Johns Hopkins University Press.

Witt, Susan, and Jay Rossier. 2000. "A New Lease on Farm Land: Assuring a Future for Farming in the Northeast." Retrieved March 31, 2006 from http://www.schumachersociety.org.

Wolf, Patricia. 2004. "The History of the Interfaith Center for Corporate Responsibility and its Place within the Wider SRI Scene in the United States." Retrieved March 31, 2006 from http://www.casi.org.

Woodhouse, Edward. 2006. "Nanoscience, Green Chemistry, and the Privileged Position of Science." In *The New Political Sociology of Science*, ed. Scott Frickel and Kelly Moore. University of Wisconsin Press.

Woodhouse, Edward, and Steve Breyman. 2005. "Green Chemistry as Social Movement?" *Science, Technology, and Human Values* 30 (2): 199–222.

Woodhouse, Edward, David Hess, Steve Breyman, and Brian Martin. 2002. "Science Studies and Activism: Possibilities and Problems for Reconstructivist Agendas." *Social Studies of Science* 32 (2): 297–319.

Worldcarfree.net. 2006. "Worldcarfree.net." Retrieved March 31, 2006 from http://www.worldcarfree.net.

World Commission on Environment and Development. 1987. *Our Common Future.* Oxford University Press.

World Council of Credit Unions. 2004. "2004 Statistical Report." Retrieved March 31, 2006 from http://www.woccu.org.

Wynne, Brian. 1994. "Public Understanding of Science." In *Handbook of Science and Technology Studies*, ed. Sheila Jasanoff, Gerald Markle, James Peterson, and Trevor Pinch. Sage.

Wynne, Brian. 1996a. "May the Sheep Safely Graze? A Reflexive View of the Expert-Lay Knowledge Divide." In *Risk, Environment and Modernity*, ed. Scott Lash, Bronislaw Szerszynski, and Brian Wynne. Sage.

Wynne, Brian. 1996b. "Misunderstood Misunderstandings: Social Identities and the Public Uptake of Science." In *Misunderstanding Science?* ed. Alain Irwin and Brian Wynne. Cambridge University Press.

Wysocki, Jr., Bernard. 2006. "Bankers Struggle to Contain Growth of Credit Unions." *Wall Street Journal*, March 7.

Yearley, Steven. 1992. "Green Ambivalence about Science." *British Journal of Sociology* 43 (4): 511–532.

Yearley, Steven. 1996. *Sociology, Environment, Globalization: Reinventing the Globe*. Sage.

Yunus, Muhammad. 1994. *Banker to the Poor: Microlending and the Battle Against World Poverty*. Public Affairs.

Zadek, Simon. 2001. *The Civil Corporation: The New Economy of Corporate Citizenship*. Earthscan.

Zald, Mayer and Roberta Ash. 1966. "Social Movement Organizations: Growth, Decay, and Change." *Social Forces* 44 (3): 327–340.

Zavestoski, Stephen, Phil Brown, Meadow Linder, Sabrina McCormick, and Brian Mayer. 2002. "Science, Policy, Activism, and War: Defining the Health of Gulf War Veterans." *Science, Technology, and Human Values* 27 (2): 171–205.

Zimmerman, Martin. 2001. "Is New Urbanism Growing Old? Andres Duany Holds Forth on Smart Growth." *Planning* 67 (6): 10–13.

Urban and Industrial Environments: The Series

Steve Lerner, *Diamond: A Struggle for Environmental Justice in Louisiana's Chemical Corridor*

Jason Corburn, *Street Science: Community Knowledge and Environmental Health Justice*

Peggy F. Barlett, ed., *Urban Place: Reconnecting with the Natural World*

David Naguib Pellow and Robert J. Brulle, eds., *Power, Justice, and the Environment: A Critical Appraisal of the Environmental Justice Movement*

Eran Ben-Joseph, *The Code of the City: Standards and the Hidden Language of Place Making*

Nancy J. Myers and Carolyn Raffensperger, eds., *Precautionary Tools for Reshaping Environmental Policy*

Kelly Sims Gallagher, *China Shifts Gears: Automakers, Oil, Pollution, and Development*

Kerry H. Whiteside, *Precautionary Politics: Principle and Practice in Confronting Environmental Risk*

Ronald Sandler and Phaedra C. Pezzullo, eds., *Environmental Justice and Environmentalism: The Social Justice Challenge to the Environmental Movement*

Julie Sze, *Noxious New York: The Racial Politics of Urban Health and Environmental Justice*

Robert D. Bullard, ed., *Growing Smarter: Achieving Livable Communities, Environmental Justice, and Regional Equity*

Ann Rappaport and Sarah Hammond Creighton, *Cooling the Ivory Tower: University Leadership to Slow Global Warming*

Michael Egan, *The Science of Survival: Barry Commoner and the Remaking of American Environmentalism*

David J. Hess, *Alternative Pathways in Science and Industry: Activism, Innovation, and the Environment in an Era of Globalization*

Index